Networked Publics

Networked Publics

edited by Kazys Varnelis

With contributions by researchers in the Networked Publics Research Group, the Annenberg Center for Communication at the University of Southern California: Walter Baer, François Bar, Anne Friedberg, Shahram Ghandeharizadeh, Mizuko Ito, Mark E. Kann, Merlyna Lim, Fernando Ordonez, Todd Richmond, Adrienne Russell, Marc Tuters, Kazys Varnelis

The MIT Press
Cambridge, Massachusetts
London, England

MIT Press books may be purchased at special quantity discounts for business or sales promotional use. For information, please email special_sales@mitpress.mit.edu.

This book was set in Garamond 3 by Graphic Composition, Inc. and was printed and bound in the United States of America.

Library of Congress Cataloging-in-Publication Data

Networked publics / edited by Kazys Varnelis.
p. cm.
Product of a fellowship program at the Annenberg Center for Communication at the University of Southern California, 2005–2006.
Includes bibliographical references and index.
ISBN 978-0-262-22085-9 (hardcover : alk. paper)
1. Internet—Social aspects—United States. 2. Internet—Political aspects—United States. 3. Online social networks—United States. 4. Convergence (Telecommunication) I. Varnelis, Kazys.
HM851.N4765 2008
303.48'330973—dc22

2008005365

10 9 8 7 6 5 4 3 2 1

Contents

Preface

Networked Publics is the product of the 2005–2006 research year at the University of Southern California's Annenberg Center for Communication. A team of thirteen scholars spent the year investigating how new and maturing networking technologies are reconfiguring the way we interact with content, media sources, other individuals and groups, and the world that surrounds us.

This book accompanies the blog that we maintained during that period, now at http://networkedpublics.org. It aims to be a scholarly introduction to the field, synthesizing our own fields of research together with what we learned as a group. The book should also be understood as our response to the cultural material and debates that we brought together at the Networked Publics Conference and Media Festival at the Annenberg Center for Communication on April 28 and 29, 2006.

Networked Publics was produced online, using tools such as Writely (now Google Docs) and, as such, is one of the first books to be produced through collaborative software.

Acknowledgments

All of us would like to thank the University of Southern California's Annenberg Center for Communication for sponsoring the Networked Publics research year. The editor is especially grateful to Dean Mark Wigley and Columbia University's Graduate School of Architecture, Planning, and Preservation for giving precious time to edit the manuscript and write the conclusion. We also give thanks to the MIT Press and our editor, Douglas Sery, for believing in us and publishing this book. Special thanks to MIT Press's Jessica Hosman for her keen eye, sharp intelligence, and good spirits. At the Annenberg Center for Communication, we would like to thank Directors Elizabeth Daley, under whose tenure this project was launched, and Jonathan Aronson, who oversaw its completion, as well as Todd Richmond, who was Managing Director during the project's inception. For their help in making *Networked Publics* happen, we are indebted to Annenberg Center staff Steve Adcock, Josie Acosta, Chris Badua, Michael Goay, Claudia Gonzalez, JoAnn Hanley, Elizabeth Harmon, Lara Mazzoni, Bryan Schneider, and LaJuana Whitner.

We would like to express our appreciation to the distinguished lecturers who provided us with their insight: Mike Liebhold, Chris Anderson, Yochai Benkler, Geoffrey C. Bowker, Howard Rheingold, and Saskia Sassen. At the media festival and conference, guest curators Steve Anderson, Jonathan "Inertia" Cullinane, Rachel Cody, Paul Marino, and Aram Sinnreich together with speakers Sean Bonner, Farai Chideya, Harry Cleaver, Zadi Diaz, Iman Foroutan, Mark Frauenfelder, James Fishkin, Tracy Fullerton, Adam Greenfield, Joi Ito, Xeni Jardin, Jonathan McIntosh, Michael Nitsche, Nicholas Nova, Mike Outmesguine, Ellen Seiter, Paul Symczak, Mark Shepard, Holly Willis, Simon Wilkie, Jason Wishnow, and Zalas generously gave their time to make the media festival possible. Many visitors graced Networked Publics over the year that our seminars took place to encourage us and to add to the discussion, among them John Seely Brown, Pierre de Vries, Henry Jenkins, Michael Kleeman, Geert Lovink, Richard Seyler-Ling, Pal Sorgaard, Tanja Storsul, and Bob Stein.

The editor would like to thank all of the members of Networked Publics, Walter Baer, François Bar, Julian Bleeker, Anne Friedberg, Shahram Ghandeharizadeh, Mizuko Ito, Mark E. Kann, Merlyna Lim, Fernando Ordonez, Todd Richmond, Adrienne Russell, and Marc Tuters. In particular, he would like to single out the editors of the sections for their work and, above all, Mizuko Ito, who not only contributed to the book but nurtured the project every step of the way. He would also like to thank Derek Lindner, Lev Manovich, and Tim Ventimiglia for their crucial readings of the conclusion, as well as the peer reviewers for MIT Press, whose comments were invaluable.

Finally, we would all like to thank our families and children for putting up with us as we took on *Networked Publics.* It is to them, and to the staff of the Annenberg Center, that we dedicate this project.

Introduction

Mizuko Ito

The era of digital media and networking is no longer in its infancy. The nineties were a pivotal decade in which computer networking became a core player in communications and media content delivery. During this period, the dominant metaphors for information technology changed from computing and artificial intelligence to networking and communication, and multimedia production and playback capabilities became standard for personal computers. With the advent of graphical browsers, the establishment of consumer broadband Internet providers, and the popular adoption of the internet, computer networking expanded its reach beyond hobbyist, research, and government communities to the broader public. Although the dot-com boom and bust of the late nineties absorbed public discourse surrounding the Internet, both during this period and afterwards, the Internet became the backbone for more and more of our everyday communications, commerce, and content delivery. At the same time, mobile phone technology became more ubiquitous and is now one of the most widely used portals to information technology. More recently, dynamic visual media such as videos and movies became available on the Internet, and increasingly sophisticated infrastructures for social exchange have heralded what some technologists are now calling Web 2.0.[1]

These technological changes are tied to important shifts in society and culture. Networked digital media are beginning to be taken for granted in everyday life. Although the nature of adoption varies widely by factors such as nation, region, class, and gender, an increasing number of people are domesticating networked digital media for their ongoing business, for socialization, and for cultural exchange. This is particularly true of the current generation of teens and young adults in postindustrial countries growing up with networked digital media as a fact of life. In Japan young people use their multimedia mobile phone as their primary communications portal, in the Philippines text messaging has revolutionized political mobilization,[2] while in Korea total penetration of homes by broadband Internet has enabled radically new forms of online sociality.

Our focus in this book, however, is on the United States, which is both an unusual and exemplary case. Although the United States has lagged behind other industrial countries in certain areas, such as the adoption of big broadband or mobile Internet, it continues to have a leadership role in the development of Internet standards, communications software, and related social practices, most recently those embedded within so-called social software. Many of the contributors to this volume have done research in other countries, and our work is informed by an international perspective; however we made the decision early on that we would focus our efforts on the national context, where we had the most collective expertise. We are not blind to the fact that other high-tech and developing countries, leapfrogging their way to wireless Internet, could be considered the cutting edge of contemporary network society and culture. Rather, our aim is deliberately parochialized to the specifics of the U.S. context, with its particular blend of concerns surrounding infrastructural development, political expression, intellectual property issues, and certain modes of cultural production. We consider the U.S. case as a specific, but still broadly influential, context of networked society and culture.

This introduction provides a preview and framework for the chapters to follow. After introducing the background to this book and the conceptual approach, I introduce themes that cut across the individual chapters by describing four key trends: accessibility to digital tools and networks, many-to-many and peer-to-peer forms of distribution, value at the edges, and aggregation of culture and information. These themes are intended as a guiding framework for understanding the relation between the chapters in the body of the book, organized by the topics of place, culture, infrastructure, and politics.

Framework

The term *networked publics* references a linked set of social, cultural, and technological developments that have accompanied the growing engagement with digitally networked media. The Internet has not completely changed the media's role in society: mass media, or one-to-many communications, continue to cater to a wide arena of cultural life. What has changed are the ways in which people are networked and mobilized with and through media. The term *networked publics* is an alternative to terms such as *audience* or *consumer.* Rather than assume that everyday media engagement is passive or consumptive, the term *publics* foregrounds a more engaged stance. Networked publics takes this further; now publics are communicating more and more through complex networks that are bottom-up, top-down, as well as side-to-side. Publics can be reactors, (re)makers and (re)distributors, engaging in shared culture and knowledge through discourse and social exchange as well as through acts of media reception. With the growth of multimedia on the Internet, publics can traffic in both professional and personal media, in new forms of communication that often find a route around commercial media distribution. Personal media and communications technologies such as telephony, e-mail, text messaging, and everyday photography and journaling are colliding with commercial and mass media such as television, film, and commercial music. This is what Henry Jenkins has described as "convergence culture, where old and new media intersect, where grassroots and corporate media collide, where the power of the media producer and the media consumer interact in unpredictable ways."[3] This book describes the current state of networked publics at the layers of place, culture, politics, and infrastructure, examining historical context and speculating about an unfolding future.

If networked media ecologies are maturing and becoming more established in our everyday lives, we are also still clearly in a moment of transition. We write this book not only to describe emergent developments in networked society, technology, and culture, but also to provide an accessible text to inform debate about our media future. For example, the chapters in the body of the book take on issues such as privacy in the rise of the Internet of things, debates over net neutrality, controversies over intellectual property in the culture industries, and whether Internet culture supports democracy and deliberative discourse.

Our method is interdisciplinary, syncretic, and collaborative. This book is a result of a year-long fellowship program at the Annenberg Center for Communication at the University of Southern California, where scholars from a wide variety of backgrounds and disciplines convened to consider the present and future of networked society and culture. From the outset, we decided against producing an edited work that would be simply a record of our diverse interests, with each scholar contributing a chapter in the well-established mode of an edited collection. To this end, we turned to the technologies that we are researching as vehicles for developing a collective intelligence. These included wikis, blogs, content management systems, and networked writing sites, as well as the usual toolkit of e-mail, instant messaging, and face-to-face and telephone conversation. A record of our work can be found at http://www.networkedpublics.org. A collaborative writing project, this book has pushed each of us beyond our specific research projects to consider the relationships between our different areas of study, working to build conceptual linkages that outline the contours of contemporary networked society in broad terms. To survey the spread of networked digital culture, it was necessary to sample areas and theoretical perspectives well beyond the comfort zone of an individual scholar. Despite the diversity of approaches that we take in this book, we share a collective commitment to an interdisciplinary understanding of sociotechnical change. The authors gathered here come from backgrounds as varied as engineering, architecture, critical studies, political science, communications, history, anthropology, and media arts. Working together demanded that we recognize the importance of a wide variety of factors including behavior, economy, culture, politics, and technology.

When writing about new technologies, it is tempting to focus on the technologies as the site of interest and the most decisive driver of change; however in this book we work actively against a technically determinist frame. One of the primary theoretical innovations of contemporary technology studies has been the recognition that technology does not stand apart as an external force that impacts society and culture. Rather, technologies are embodiments of social and cultural structures that in turn get taken up in new ways by existing social groups and cultural categories.[4] As Lawrence Lessig famously argues in the case of legal structures being embodied in technical architectures, "law is code."[5] Similarly, John Seely Brown and Paul Duguid have argued that information has a "social life" that structures its uptake and creation.[6] This stance is foundational to the interdisciplinary approach we take. Sociocultural factors are subject to technical analysis just as technical factors are subject to social and cultural analysis. This stance also demands that our writing is not focused on specific new technologies, but rather on longstanding social, cultural, technical, and material domains. The chapter topics—place, culture, politics, and infrastructure—are meant to locate contemporary technologies within broader historical trajectories.

This recognition of the social, culture, and material nature of information technology is not only a research commitment; it is also a sign of the technological times. As computers have moved from being standalone boxes that were "computing machines" or "models of the mind" to being networked devices for human communication, our popular understandings of computers have also changed. We look to the online world as a source of sociality and culture, and designers of new online systems recognize that they are engaged in social engineering as well as technical engineering. For many, computers and digital technologies have become intimate, indispensable, and pervasive in their lives. More recently, with the advent of portable networked technologies such as the mobile phone and RFID (Radio Frequency Identification) tags, as well as location-based networked systems, we are also being forced to recognize information systems' relation to the materiality of diverse objects and places. In other words, the interdisciplinary approach that we take in this book is tuned to the current moment in networked culture and society, a moment when we are actively grappling with the massive convergence of society, culture, places, and things via the medium of the Internet.

Our review of networked society and culture is certainly not comprehensive. We do not, for the most part, delve into the details of particular technologies, platforms, or online sites. The reader should not expect coverage of many of the developments in online systems that have been most visible in public discourse, whether that is multiplayer online games, eBay, blogs, MySpace, or YouTube. Our intention is not to rehash material already covered by the popular media as well as a growing body of scholarly case studies. Instead, we direct our efforts toward reading across these different developments to identify broader patterns and shifts in culture and society. We mobilize case studies that speak to these broader trends, though they may not be the trends that are most visible or debated in public culture today. This book is also not intended as a theoretical text that proposes a new framework for understanding digital networks. In recent years, significant texts have been published that are defining the terms of debate in this area. We introduce what we believe to be key thinkers and concepts for understanding networked and convergent culture and society. Recent theoretical texts that have been particularly influential to our enterprise are featured in boxes throughout this book. Our goal is to bring these theoretical contributions into conversation with one another and into relation with the wide range of content areas and examples that we have collectively researched. In this way, we have tailored this book to the strengths of a collective enterprise and collaborative writing process.

I now turn to an overview of some overarching themes covered throughout the book: the accessibility to digital production and networking tools, peer-to-peer and many-to-many forms of content distribution and publishing, value at the edges, and aggregation of information and culture.

Accessibility

The current growth of networked publics is grounded in the spread of digital technologies and networks. Lowered costs of processing power and digital storage, accessibility of various digital production tools, as well as more pervasive network infrastructures—particularly through mobile and wireless technologies—are all important factors. Yochai Benkler characterizes this as one of the central shifts toward a networked information economy: "the move to a communications environment built on cheap processors with high computation capabilities, interconnected in a pervasive network."[7] This distribution of processing power to larger masses of people is linked to an unprecedented spread of the means of cultural and information production and dissemination. It has been a decade or so since access to the production of text-based digital content through word processing, text messaging, and e-mail became relatively common in the United States. More recently, easy-to-use Web editors, blogging software, and digital cameras enabled multimedia publication using a standard personal computer toolkit. Now with the spread of digital video cameras, the means to produce video are readily available as a standard package of personal computer functionality. Software programs such as iMovie and GarageBand, and Web sites like Flickr and YouTube, are exemplary of this new ability in everyday life to author rich digital media. Taking this idea further, the chapter on culture describes the growth of amateur digital content being shared online, and the growing salience of cultural styles of remix and appropriation. Similarly, the chapter on politics describes new modes of bottom-up political expression and mobilization that are enabled when the means of digital production are close at hand.

In addition to the distribution of the means of cultural and knowledge production, networking infrastructures are becoming increasingly pervasive and varied. The intimate presence of the mobile phone in our everyday lives is probably the most emblematic shift in this relation of network accessibility. Users rely on handheld devices to maintain an always-on relation to information and personal networks, as well as utilizing them as ready-at-hand digital production devices for snapping photos and crafting text messages. In addition, the presence of Wi-Fi and other wireless Internet infrastructures is growing, along with experimental efforts in Internet-connected automobiles and location-based networking services. The chapter on place describes how pervasive digital networks are reconfiguring our relation to place by enabling simultaneous presence in both physical and networked place. This layer of networked accessibility is tied to a range of social and cultural tensions—drivers are distracted by their mobile phones and screens; massive, multiplayer online games capture players' attention at the expense of out-of-game commitments; parents and children alike text others from the dining room table; and people congregate in cafés only to huddle in front of their laptops. We are still very much in the midst of negotiating appropriate social norms in this era of layered presence.

The issue of pervasive networked connectivity involves the politics of objects and infrastructure as well as interpersonal social negotiations. The chapter on place describes how locations and objects are becoming part of networked publics through technologies such as Geographic Information Systems and

RFID tags. As these new systems are deployed to map the traffic in objects and the characteristics of places, we should expect to see a new set of social controversies about privacy and the invasiveness of digital networks. The chapter on infrastructure sounds another cautionary note, warning us not to assume that networking infrastructures are always deployed in even and equitable ways. This chapter describes the policies and politics surrounding the deployment of "big broadband," and the political and economic obstacles that stand in the way of cheap, accessible broadband in the United States. The digital divide is resilient because the bar of technological sophistication continues to rise. Even as larger masses of people gain access to digital technologies and networks through mobile phones, big broadband and state-of-the-art personal computers remain out of reach for most.

Peer-to-Peer and Many-to-Many

From the growing accessibility of digital tools and networks have come new means and practices for distributing digital content. As the chapter on infrastructure notes, from its inception the Internet has relied on an open end-to-end (E2E) architecture that has prioritized the free flow of content from the ends, rather than being selective about types of content or where the content traveled. As the Internet has scaled up, and as networking applications have become more sophisticated, this E2E architecture has helped support cultures of *peer-to-peer* (P2P) media distribution and *many-to-many* (M2M) forms of communication.

From the eighteenth century to the present day, media and knowledge were largely compartmentalized into either interpersonal talk and dialog or the mass copying and physical distribution of objects such as paper, tapes, CDs, and DVDs. On one end of the spectrum, large-scale media distribution was controlled by commercial industries and their one-to-many infrastructure of broadcast and commodity distribution. On the other end, personal communication was dominated by one-on-one or small-group talk through modalities such as physical gatherings and telephone conversations. Local and amateur media existed in the form of pamphlets, zines, and community media of various kinds, but access to these media forms was limited in terms of resources and reach. P2P Internet tools enable the M2M distribution of amateur and niche content as well as the one-to-many distribution more characteristic of commercial media.

Anyone with access to an Internet connection has a soapbox with which to try and reach their audience, even if that audience is spatially dispersed. P2P distribution systems such as Napster, Kazaa, and BitTorrent, M2M sharing platforms such as DeviantArt, Flickr, Fanfiction.net, and YouTube, and social networking tools such as MySpace, LiveJournal, and Facebook radically expand opportunities for individuals to share media and information directly with others in a social context. With sites such as eBay, Amazon.com, Lulu, Etsy, and Yahoo! Auctions, tangible objects flow through P2P networks, spurring new forms of microenterprises built on secondary market exchange. All of these sites also function as content aggregators that enable niche creators and specialized audiences to find one another. This new mode of M2M distribution has resulted in what Chris Anderson has described as the *long tail* of media distribution, where sites like Amazon.com are increasingly making money from the small sales of large numbers of niche products rather than just massive sales of bestsellers.[8] Aggregation of M2M distribution also means that media content that may have started in a niche has the potential to reach massive audiences, as we've seen in the cases of the Drudge Report, Red vs. Blue, or YouTube celebrities.

Yochai Benkler sees these decentralized networks of communication and exchange as major catalysts of the shift to a networked information economy that is displacing the industrial information economy. In this economic model, "decentralized individual action—specifically, new and important cooperative and coordinated action carried out through radically distributed, nonmarket mechanisms that do not depend on proprietary strategies—plays a much greater role."[9] In a similar vein, Michael Bauwens sees P2P as an increasingly salient form of human dynamic that is social, economic, and political in nature, and goes so far as to suggest that it is the "premise of the next civilizational stage. . . . It's a form of human network-based organisation which rests upon the free participation of equipotent partners, engaged in the production of common resources, without recourse to monetary compensation as key motivating factor, and not organized according to hierarchical methods of command and control."[10] As both Benkler and Bauwens suggest, lowered barriers to the means of distribution have meant that the reach of nonmarket sharing of knowledge and culture has expanded dramatically.

In the early days of the Internet, Howard Rheingold described a culture of "virtual community," characterized by supportive interpersonal interaction.[11] This culture of community-based sharing is still very much alive in many corners of the Internet—in LiveJournal communities, online game guilds, MySpace networks, mailing lists, and Yahoo! groups. But these interpersonal networks have been radically augmented by sharing between relative strangers mediated by new sociotechnical systems. People provide content free and anonymously to others via P2P systems such as BitTorrent or Kazaa, muddying the boundaries between what some see as sharing and others have labeled piracy.[12] One-time visitors to these interpersonal networks scatter comments on blogs, and anonymous others browse and comment about personal photos on sites such as Textamerica and Flickr. Social network and reputation systems on eBay and Amazon.com, Technorati link tracking for blogs, and comment "karma" on sites like Slashdot help us navigate and prioritize the massive marketplace for M2M exchange.

P2P networks have different dynamics in the spread of information, tending toward more viral word-of-mouth circulation rather than top-down dissemination. As the chapter on culture describes, a wide range of players have exploited this—including activists packaging political messages in catchy videos, established commercial media using blogs and reader participation, marketers employing these same techniques in viral advertising campaigns, and bands and filmmakers making use of promotional sites to generate buzz through P2P networks. Not surprisingly, there has been a backlash to *astroturfing,* or public relations campaigns that attempt to simulate grassroots behavior. A text message, or Short Message Service (SMS), zapped from friend to friend has proven to be a potent tool for organizing spontaneous political protests as well as more playful gatherings of flashmobbers. Technologists are also exploring the potential to use these P2P dynamics to design wireless-mesh networks that rely on relaying network traffic between individual users rather than a centrally managed network. These viral models are described respectively in the chapters on politics, place, culture, and infrastructure.

The growth of P2P traffic in commercial content has led to a wide range of opportunities as well as new social problems. The most high-profile battles have been with respect to P2P exchange of commercial works such as music, television, and movies. Culture industries are struggling to regulate and monetize the traffic of their content over P2P networks. This has led to high-stakes battles over intellectual property policy and digital rights management technologies.[13] Some of these dynamics are described in the chapter on networked public culture. The underlying issue is the tension between openness and control in the flow of culture and information. In an E2E environment, people also begin to see value in filtering, regulating, and prioritizing the flow of content. This tension appears in the debate over network neutrality, as described in the infrastructure chapter. Commercial content providers are beginning to explore alliances with Internet service providers to filter network traffic in order to prioritize commercial content delivery of P2P traffic. A similar tension is at work in the domain of politics. Although the Internet has spurred a rise in online political discourse, it has been difficult to channel these conversations in ways that conform to the norms of productive political deliberation. The chapter on politics describes the struggle of political activists and theorists to foster political deliberation.

Value at the Edges

In their discussion of business strategy in an era of globalization, John Hagel and John Seely Brown suggest that we should increasingly look to the edges—the edges of companies, markets, geographies, and demographics—to find innovation. They look to the E2E architecture of the Internet as both an enabler and a metaphor for value creation at the edges.[14] The current growth of activism at the ends of the network and media ecology has implications for a wide range of social, cultural, and economic domains. For example, in politics, the Internet has led to a growing visibility of smaller-edge political actors who can make their voices heard in political blogs, make small campaign contributions, coordinate events via viral SMS exchange, or mobilize supporters through networked activist groups. In a similar vein, the chapter on networked public culture focuses on the changing relationship between media producers and consumers, describing the cases of the industries for music, anime, advertising, and news. The growing activism of media audiences in, what Jenkins has called, a "participatory media culture"[15] reverberates back to media industries, reconfiguring the relationship between the edge and the core. The result is new configurations of media markets characterized by proliferating special-interest groups that dwarf what was previously considered the mainstream. This is the core of what Chris Anderson has described as the long-tail phenomenon in media markets.[16]

With an expanded network, individuals are able to reach out to a potentially larger and more varied pool of culture and information. While debates on globalization in the heyday of mass media suggested that interconnection would lead to the homogenization of culture, in the Internet era the opposite appears to be the case. What we are seeing now is a proliferation of niches in subcultures, such as English-language fandoms of Japanese animation, a case described in the culture chapter. Teens anywhere in the United States can gain access to niche media from Japan that they would never have been able to get their hands on even a decade ago. Nevertheless, in the blogosphere, this tendency has been criticized as creating an echo chamber: bloggers and audiences are connecting, at greater frequency and fidelity, with people who share their opinions, relying less on the standards of neutrality espoused by the mainstream press. At a lower level of granularity, we also see this in the *telecocoons*[17] described in the chapter on place. Mobile phones, Wi-Fi hot spots, and networked automobiles create personal cocoons of private connectivity and conversation so people can stay connected with the people they feel most comfortable with. At the same time, these technologies have also been criticized as leading to social insularity, as people shut out engagement with copresent others in favor of their remote, but intimate, relations.

This tendency toward niches, peer cultures, and special-interest groups has been widely criticized as leading toward a fragmentation in common culture and standards of knowledge. This critique has been particularly noticeable in the case of news, where professional journalists worry about the breakdown of civic culture and journalistic standards as people turn to the blogosphere for news and opinion. This case is described in the culture chapter. Similarly, the politics chapter describes how deliberative democrats worry that online political discourse is just chatter and is rarely elevated to the level of true deliberation that can have political clout. Within infrastructure policy, this concern manifests in the griping of network providers who bear the expense of wiring the communications backbone. These providers feel that they are providing services for the edges without being able to recoup revenue back to the core. Commercial content providers have a similar complaint—they feel they are providing investment and value into cultural resources that get exploited by the edges without the circulation of revenue back into the industry. Commercial content and network infrastructure providers are joining hands to lobby for the rights of the core to regulate and capitalize on the value being trafficked at the edges, through digital rights management schemes and filtering and by the prioritizing of network traffic.

Aggregation

As networks expand, the dynamic tension between the broader network and individualized niches becomes more pronounced. This is a dynamic that Manuel Castells has famously dubbed the relation between the network and the self,[18] a relation that, in the conclusion of this book, Kazys Varnelis suggests is undergoing a fundamental shift. The growth of value at the edges is linked to the aggregation of a growing range of media content through the Internet and various mechanisms of searching and filtering. As the Internet has evolved from a medium for the exchange of text, to include pictures, sounds, video, and 3D worlds, the scope of the culture and knowledge that is available for digital aggregation and access has expanded dramatically. We not only exchange text and pictures to family and friends, but also links to video and other rich media. Search companies like Google and Yahoo! are busily constructing new systems to aggregate and filter video, particularly after YouTube entered the public eye. Game systems like Second Life are putting forth three-dimensional online worlds as general-purpose interfaces to knowledge, culture, and sociability, hoping to renew the desire for a three-dimensional metaverse ignited by cyberpunk authors such as William Gibson and Neal Stephenson.[19] Further, as described in the chapter on place, the embedding of networks in location, the creation of networks of location (as in the case of Google Earth), and the possibility of an Internet of things[20] is also on the horizon. Objects and places are the next targets for aggregation into the digital network. As networks increasingly pervade the nooks and crannies of physical space through portable objects and place-based infrastructure, we have opportunities for an always-on sense of networked connectivity and a layering of presence in various physical and online places.

The effects of this large-scale aggregation of knowledge and culture are varied and often contradictory. Scholars such as Jenkins and Benkler note more powerful and distributed collective intelligences that are enabled by new networking systems. Benkler describes a wide range of cases such as Wikipedia and SETI@home as instances of what he calls "commons-based peer production": "radically decentralized, collaborative, and nonproprietary; based on sharing resources and outputs among widely distributed, loosely connected individuals who cooperate with each other without relying on either market signals or managerial commands."[21] Jenkins suggests that "consumption has become a collective process"[22] and describes examples of highly mobilized fan and gaming groups who develop vast stores of collective knowledge about their hobbies. Political groups such as MoveOn.org suggest a new model of leveraging network aggregation for political mobilization.

This aggregation effect in the nonmarket sector also has a counterpart in the commercial sector. Anderson's analysis of the long tail describes the proliferation of niches as tied to the business success of network-based content aggregator sites such as Amazon.com or Rhapsody.[23] Niches feed the aggregators and vice versa in a cycle either virtuous or vicious, depending on your perspective. The advent of sophisticated recommendation and reputation systems also feeds the tendency of individuals to rely on aggregator sites as channels to niche interests and products. Although network participants may not rely on centralized sources of knowledge such as mainstream news sites or information portal sites, they increasingly turn to search engines and aggregated service sites. The presence of Google as a new information industry behemoth with unprecedented power is testament to the power of aggregation services at this current moment in network society. Network aggregation is taking new forms as objects and locations become integrated into digital networks. The chapter on place describes current speculation about the role of everyday objects as they become increasingly networked. The incorporation of geographic information systems into our everyday lives is well under way with the advent of services such as Google Earth and MapQuest. Now aggregated geodata is accessible to corporations and individuals, allowing companies such as Claritas .com to map the relation between places and lifestyle demographics, or Zillow .com to sell real estate by letting individuals fly in virtual space over neighborhoods mapped to property values. These systems highlight the ways in which aggregation often crosses certain boundaries of privacy, boundaries that are likely to become more pronounced if objects are also transmitting information to the network. Bloggers, webcams, and camera phones now upload a steady stream of information to the Internet—information that can be easily searched, tagged, and reblogged. Already we are seeing a series of moral panics surrounding privacy and accessibility to personal information on sites such as MySpace or Facebook, a louder echo of earlier social problems when search engines first began crawling Net newsgroups and mailing-list archives.

With this we return full circle to the issue of accessibility and ubiquity of the network. The four themes I have outlined—accessibility, P2P and M2M distribution, value at the edges, and aggregation—are threads woven throughout the chapters to follow. The chapters address these technosocial trends in more depth, within specific domains and case studies. Collectively, they trace the contours of an emerging set of networked publics, describing their historical evolution and suggesting the current controversies that are likely to shape their future.

Notes

1. Tim O'Reilly, "What Is Web 2.0: Design Patterns and Business Models for the Next Generation of Software," *O'Reilly Network,* http://www.oreillynet.com/pub/a/oreilly/tim/news/2005/09/30/what-is-web-20.html.

2. Mizuko Ito, Daisuke Okabe, and Misa Matsuda, eds., *Personal, Portable, Pedestrian: Mobile Phones in Japanese Life* (Cambridge, MA: MIT Press, 2005); Jon Agar, *Constant Touch: A Global History of the Mobile Phone* (Cambridge, UK: Icon Books, 2003); Howard Rheingold, *Smart Mobs: The Next Social Revolution* (Cambridge, MA: Perseus, 2002).

3. Jenkins, *Convergence Culture: Where Old and New Media Collide* (New York: New York University Press, 2006), 2.

4. Paul Edwards, "From 'Impact' to Social Process: Computers in Society and Culture," in *Handbook of Science and Technology Studies,* ed. Shelia Jasanoff, et al. (London: Sage, 1995); Christine Hine, *Virtual Ethnography* (London: Sage, 2000); Trevor F. Pinch and Wiebe E. Bijker, "The Social Construction of Facts and Artifacts: Or How the Sociology of Science and the Sociology of Technology Might Benefit Each Other," in *The Social Construction of Technological Systems,* ed. Wiebe E. Bijker, Thomas P. Hughes, and Trevor Pinch (Cambridge, MA: MIT Press, 1987).

5. Lessig, *Code and Other Laws of Cyberspace* (New York: Basic Books, 1999).

6. Brown and Duguid, *The Social Life of Information* (Boston: Harvard Business School Press, 2002).

7. Benkler, *The Wealth of Networks: How Social Production Transforms Markets and Freedom* (New Haven: Yale University Press, 2006), 3.

8. Anderson, *The Long Tail: Why the Future of Business Is Selling Less of More* (New York: Hyperion, 2006).

9. Benkler, *The Wealth of Networks,* 3.

10. Bauwens, "Peer to Peer and Human Evolution," Integral Visioning, http://integralvisioning.org/article.php?story=p2ptheory1.

11. Rheingold, *The Virtual Community: Homesteading on the Electronic Frontier* (New York: Addison Wesley, 1993).

12. Shuddhabrata Sengupta, "A Letter to the Commons," in *In the Shade of the Commons: Towards a Culture of Open Networks,* ed. Lipika Bansai, Paul Keller, and Geert Lovink (Amsterdam: The Waag Society, 2006).

13. See Lawrence Lessig, *Free Culture: How Big Media Uses Technology and the Law to Lock Down Culture and Control Creativity* (New York: Penguin, 2004).

14. Hagel and Brown, *The Only Sustainable Edge: Why Business Strategy Depends on Productive Friction and Dynamic Specialization* (Cambridge, MA: Harvard Business School Press, 2005).

15. Henry Jenkins, *Textual Poachers: Television Fans and Participatory Culture* (New York: Routledge, 1992); Jenkins, *Convergence Culture.*

16. Anderson, *The Long Tail.*

17. Ichiyo Habuchi, "Accelerating Reflexivity," in *Personal, Portable, Pedestrian: Mobile Phones in Japanese Life,* ed. Mizuko Ito, Daisuke Okabe, and Misa Matsuda (Cambridge, MA: MIT Press, 2005).

18. Castells, *The Rise of the Network Society* (Cambridge, MA: Blackwell, 1996).

19. Gibson, *Neuromancer* (London: Victor Golancz, 1985); Stephenson, *Snow Crash* (New York: Bantam Books, 1992).

20. International Telecommunication Union, "ITU Internet Report 2005: The Internet of Things," (Geneva: International Telecommunication Union, 2005), 2–5.

21. Benkler, *The Wealth of Networks,* 60.

22. Jenkins, *Convergence Culture,* 4.

23. Anderson, *The Long Tail.*

1

Place: The Networking of Public Space

Kazys Varnelis and Anne Friedberg

Contemporary life is dominated by the pervasiveness of the network. With the worldwide spread of the mobile phone and the growth of broadband in the developed world, technological networks are more accessible, more ubiquitous, and more mobile every day. The always-on, always-accessible network produces a broad set of changes to our concept of place, linking specific locales to a global continuum and thereby transforming our sense of proximity and distance.

In the following chapter, we explore both the networking of space and the spatiality of the network, identifying a series of key conditions: the everyday superimposition of real and virtual spaces, the development of a mobile sense of place, the emergence of popular virtual worlds, the rise of the network as a socio-spatial model, and the growing use of mapping and tracking technologies. These changes are not simply produced by technology. On the contrary, the development and practices of technology (as well as the conceptual shifts that these new technological practices produce) are thoroughly imbricated in culture, society, and politics. To be clear, the new is not good by default. The conditions we observe are contested and give rise to new tensions as much as to new opportunities. With connection there is also disconnection, and networks can consolidate power in the very act of dispersing it. We will examine both the plusses and minuses of these conditions throughout the chapter.

Taken together, these changes are already radical. But it is likely they will be only the first steps in restructuring our concept of spatiality toward a reality of which we can only be partially aware, just as the first theorists of modernism could only partially understand the emerging condition of their day.

Simultaneous Place: Networked Publics

"One hundred dollars for three thousand minutes," a twenty-five-year-old man with a Farsi accent repeats into his mobile phone. The scene is the local Starbucks, where you've gone to get away from the all-consuming distraction the Internet introduces into your life. You've intentionally left your phone in the car in order to be blissfully unaware of any professional or personal obligations that might take you away from your task. You've even left your laptop behind so that you won't be tempted by the queue of e-mails to catch up to. You're in the café with your Moleskine notebook—a non-networked object ubiquitous among the digerati—trying to start an essay on the role of place in network culture and finding that the only way forward is to detach yourself from the network as much as possible. But the people surrounding you have other ideas. The man behind you is trying to commit himself more deeply to the network, purchasing a plan that will allow him to talk on his mobile phone for one-tenth of his waking hours every month. A woman next to you is browsing the Internet with her laptop, a late-career executive is thumbing his Blackberry, two students are studying together, and some teenagers are hanging out listening to music on their iPhones. While one texts her friends, the other downloads music from the iTunes store. A thirtysomething man is on his laptop working on a screenplay, while a few people are just reading books or the paper. You are all somehow drawn together by the lure of the generic (but branded) caffeinated beverage and the desire to share a similarly generic, but nonetheless communal, space with other humans with whom you are likely not to have any direct interaction.

This is, as far as humanity goes, a scene that is simultaneously age-old and unprecedented. We gather at the communal watering hole as we always did; only now we don't reach out to those around us. Instead, we communicate with far-flung souls using means that would be indistinguishable from magic for all but our most recent ancestors.

That we open at a coffeehouse is not incidental. For theorist Jürgen Habermas, when the public sphere emerged in the early eighteenth century, it did so in the context of the café, the learned society, and the salon.[1] Together with the rituals of coffee drinking, the café increasingly provided both forum and fuel for critical debate about the latest pamphlets, newsletters, and broadsides. But the public sphere was not so much a physical place as a discursive site in which a literate public could conduct rational and critical debate. The assembly and dialogue that constituted this emerging public sphere occurred as much within the pages of newly circulated printed materials as it did within the walls of the coffeehouse. And yet, although Habermas's ideas of the emancipatory potential of the public sphere were dependent on open models of communication and participation, the café maintained its own invisible divides of power and access. The spaces that Habermas championed as the original outposts of this deliberative democracy were not open to women, or to men not of the appropriate race, class, or ethnicity. Instead, women conducted different modes of deliberative discourse in separate spaces, such as the tea table and the public laundry.

What kind of public do we have in the café scene of our quotidian present? Women sit alongside men, and the patrons vary widely in age and ethnicity. But they are not engaged in debate or dialogue with each other. If they come together, it is simply to establish an ambient visual experience of bodies in near proximity, which is as psychically necessary in this wired and wireless age as it was in the days of *Australopithecus.* The material space of Starbucks is designed to facilitate this through its neighborhood location, its anonymous yet familiar design choices, its comfortable furniture, and the carefully calibrated background music. But if these individuals don't interact with the other café-goers verbally, they are engaged in a calculated copresence: while comfortably sipping coffee or its commodified equivalent in the franchised design of this local Starbucks, they are—via a network connection, mobile phone, or wireless laptop—in another place.

Of course, much has happened since the eighteenth-century café inaugurated public life in Europe's great cities. The newspaper, penny novel, and other printed matter offered new forums for the literate public; arcades, boulevards, and other public spaces shaped the bourgeois city. Poet and art critic Charles Baudelaire championed the man on the street—the *flâneur*—who moved through this newly forged urban milieu with the privilege of a bourgeois gentleman in public space: "To be away from home, and yet to feel at home; to behold the world, to be in the midst of the world yet to remain hidden from the world."[2] In reflecting on these changing configurations of public and private on the streets of modernity, German cultural critic Walter Benjamin wrote: "The street becomes a dwelling for the *flâneur;* he is as much at home among the facades of houses as a citizen is in his four walls. . . . The walls are the desk against which he presses his notebooks; news-stands are his libraries and the terraces of cafés are the balconies from which he looks down on his household after his work is done."[3]

Not everyone had the bourgeois male's privilege. The *flâneuse,* if she was to navigate public space on her own without being thought of as a product for sale, was safest in the cathedrals of consumption, department stores that encouraged other new behaviors. Consumerism was born out of the need to create desire for the products of the industrialized machines of capital and to award women new agencies, a new "purchase" on public space.

As the nineteenth century drew to its end, the pedestrian mobility of the *flâneur* and *flâneuse* was augmented by the many machines of transport—trains, streetcars, buses, moving walkways, escalators, elevators—that not only accelerated movement but produced new social behaviors. "Before the development of buses, railroads, and trams in the nineteenth century," writes sociologist Georg Simmel, "people had never been in a position of having to look at one another for long minutes or even hours without speaking to one another."[4] Yet, as both Simmel and Baudelaire observed, the only way that humans could navigate the overwhelming condition of the metropolis was by disconnecting, by shutting off their connections to this multitude of others.

Cultural critics observed that such detachment increased during the twentieth century as people fled decaying cities to suburbs. Public space became increasingly privatized and virtualized, with networks of individuals being replaced by television broadcast networks, and individuals becoming less and less citizens and more and more consumers.[5] For these critics, it wasn't just television that produced these changes. The public sphere was being evacuated, and along with it place—as well as its deeply etched social and historical meanings—was quickly disappearing. In her book *The Death and Life of Great American Cities,* Jane Jacobs linked the decline of the city and the collapse of the public sphere, arguing that a vital sense of *civitas* depends on an architectural infrastructure that encourages frequent, random face-to-face interactions within an urban community. For Jacobs, both modern urban planning and the detached single-family house in the suburb inhibited those vibrant interactions.[6]

Perhaps the crescendo of this gloom came only a decade and a half ago when anthropologist Marc Augé made his dismal conclusion about the nature of human interaction in physical space in his *Non-Places: Introduction to an Anthropology of Supermodernity.* Augé suggests that our sense of place, as old as humanity, is coming to an end. Building on Marcel Mauss's idea of place as a "culture localized in time and space," Augé distinguishes places—locations in which individuals with distinct identities form human relationships that in turn accrete, creating the sediments of history—from non-places—spaces of transition absent of identity, human relationships, or the traces of history. Augé's non-places are in-between spaces, sites of transit for humans (airports, airplanes, freeways, parking garages, but also refugee camps and shantytowns), data (the space in front of the computer screen), and goods and capital (the space in front of the ATM, the shopping mall, the supermarket). This new world of non-place, Augé writes, privileges the fleeting, ephemeral, and contingent.

Places are filled with individual identities, language, references, and unformulated rules; non-places are spaces of solitary individuality. Much as at our

Starbucks, that anonymity is shared by many. According to Augé, we are all passengers on an airplane or drivers on a highway, our identities lost. Information in the world of non-place is conveyed through disembodied texts and voices offering prescriptive information: "No smoking in the airport," "Flight 140 to Madrid departs at gate 25," "Have your passports ready and customs forms filled out," and of course, "Please take the ticket." If modernity, Augé concludes, was still deeply tied to place and history—indeed, historical narratives were key to that moment—supermodernity leaves us in a realm finally devoid of history. To be fair, Augé issues a disclaimer, noting that non-place and place are only conceptual poles. He admits that there is no such thing as pure non-place—after all, for an aviation buff, airport worker, or true road

Box 1.1 Mobile Communications and the Public Sphere

From Richard Ling, *The Mobile Connection: The Cell Phone's Impact on Society* (San Francisco: Morgan Kaufmann, 2005), 193.

In "The Tragedy of the Commons," Garret Hardin discusses the ways in which common resources are progressively used up by the individualistic rationality of participating actors. If we consider the public sphere as a type of commons, the mobile telephone brings up two issues. The first is what we might call a type of audio pollution of the public sphere because of the increasing number of mobile telephone calls. This may be a transitory problem. Just as noise pollution from other sources was seen as a problem during certain phases of technological development, the developments themselves and our ways of dealing with them have resulted in a code of behavior that soon removed the problem. SMS, for example, is one adjustment, as are other developing forms of courtesy. Here it seems that the sense of the commons is being reasserted through various adjustments.

The other issue is the withdrawal from the public sphere. As Jane Jacobs noted, the thing that makes the public sphere vibrant is the continual contact with unexpected forms of interactions. Not all are pleasant, and not all are sought. Nonetheless, there is vitality and a roundness that arise from our interaction with a variety of others, no matter how perfunctory. Seeing the legless beggar, watching a street musician, giving directions to the tourist, and seeing the exotic hair color and shockingly mismatched clothes of the older woman are all elements that inform us as to the mood and sprit of our local situation. Better this than some Stepford-like existence in which all is neatly tucked into the same pattern and alternatives are not only frowned upon, but eradicated.

At a milder level, being part of the public sphere means that we are available to tell another passenger on the bus that this is the bus stop they asked about. It means being able to ask another what the time of day is or to comment, no matter how obliquely, on the weather. Clearly when we are in the public sphere, we are only minimally social. Nonetheless there is a social component. There is, however, the possibility that ICTs and mobile communication will take a small bite out of the already minimal sociability that is available in this sphere.

warrior, the airport has history, and a highway engineer's trained eye can identify just when an overpass was built. Nor is there pure place—all places are non-places for those who have not accumulated lived experience within them. Nevertheless, Augé concludes, our era is increasingly dominated by non-place, our existence doomed to solitude.[7]

But what of place today? To be sure, the old world of public space has not magically returned. Our Starbucks is a generic space in which many alight temporarily, not a place defined by the kind of encounters that might have occurred in Vienna's Café Central where Trotsky, Freud, and Loos came every day. Our Starbucks is not a place where random individuals chat with one another about the issues of the day, at least not usually.

But face-to-face encounters are only one level of human interaction: the Starbucks anecdote suggests that, for some reason, we still have an urge to gather together, even if in our solitude. And this idea of solitude is deceiving: a great deal has changed since Augé's day. The proliferation of mobile phones and the widespread adoption of always-on broadband Internet connections in homes and offices in the developed world means that we are not necessarily alone even if we are not interacting with those in close physical proximity to us.

On one level, what we observed in Starbucks was a generic space of anonymity, its caffeinated habitués lost in the crowd. But on another level it is a place where these individuals share their proximity with others similarly engaged in a place that is networked and elsewhere. For those who gather in these hot spots to engage with the network, being online in the presence of others is the new *place* to be, the bodily presence of the other cafégoers easing the disconnect with the local that the network creates.[8]

Mobile Place: The Rise of the Telecocoon

In *The Social Impact of the Telephone,* Ithiel de Sola Pool explains how the telephone made possible the modern city, with its concentrated downtown core and increasingly dispersed suburban sprawl. In de Sola Pool's analysis, the telephone enabled remote surveillance of the factory by managers who could then work in city cores, highly packed environments where they could meet other managers for face-to-face meetings in skyscrapers—a building type viable only once the telephone had made messenger boys obsolete. As de Sola Pool points out, this technology was successful because of its context. Telephones reshaped the city by symbiotically exacerbating certain trends; capital had accumulated and specialized to the degree that further growth demanded the concentration of managerial, service, and information industries in the city core and manufacturing on the periphery.

The telephone was a technology that both encouraged sociability and maintained intimacy at a distance. Unlike radio's monologic address to many, the telephone distributed its dialogic potential to individuals, allowing relationships to be constructed and maintained in a world marked by greater migration as well as interstate and international commerce.[9] Television, on the other hand, succeeded in the 1950s and 1960s in part because it offered a compensatory sense of belonging in rapidly expanding suburban environments (and rapidly shrinking urban environments) within which individuals already felt isolated.

As cultural theorists like Marshall McLuhan and Joshua Meyrowitz remind us, not only did television knit a global village of telepresent images by broadcasting live across its early networks, the medium produced a simultaneous doubling of place. Broadcast historian Paddy Scannell writes: "Public events now occur, simultaneously, in two different places: the place of the event itself and that in which it is watched and heard."[10]

In this regard, it is crucial to understand that humans organize space in such a way that it is a medium of its own. The city, as communications theorist Ronald F. Abler observed, is itself a communication device.[11] Until recently the two primary means of browsing this communication device have been on foot—the method of the *flâneur* or *flâneuse*—and with the automobile. The latter anticipates the condition of a mobile, networked world in that the automobile—which transports its driver and passengers in the comfort of a private interior—has always been a mobile communication device, a viewing machine; its windscreen a membrane that both protects the driver and frames the view. As automobile speed and design efficiency increased, so did a network of expanding roadways and highways—most notably the Interstate Highway System, itself a prototype for the Internet and a network of networks. As Jean Baudrillard asserted in his 1983 quip about the "private telematics" of driving, "the vehicle becomes a kind of capsule, its dashboard the brain, the surrounding landscape unfolding like a televised screen." It's no surprise the emerging metaphor for screenic access to Internet in the early 1990s was the "information superhighway."[12]

Automobiles are, in a sense, transitional mobile devices, accustoming individuals to browsing while in motion and to the experience of mobility with access. Car radios and, by the late 1970s, citizen-band radios (or CBs) connected the driver to information and communication beyond the vehicle. With mobile phones and Global Positioning System (GPS) devices providing access to a more browsable, pervasive network, we can access networked place with a wind-in-the-hair mobility—riding the train, walking down the street, sitting on the bus, or driving in car. Often, the car itself acts as an active agent. For example, systems such as General Motors's OnStar offer automatic monitoring of automobiles via special cellular transmitters and GPS units. After automatically conducting a monthly checkup on the car's health, OnStar e-mails both the automobile owner and vehicle dealer with the results and, if an airbag has been deployed, telephones the OnStar call center immediately with the vehicle's coordinates.

With the proliferation of screens in cars, as well as cellular, satellite, and Bluetooth connectivity, the driver's interface is hypermobile. Whether a car is hurtling or crawling through space, the driver's telematic connection to GPS information continually updates its location. The windshield now competes with a multitude of screens, from dash-mounted LCDs (Liquid Crystal Displays) that display navigational maps and Bluetooth phone keypads to DVD players to heads-up displays on the windshield.

Telecommunications researcher Rich Ling suggests that the flexibility the car gave to individuals and our newfound ability to coordinate with others via the mobile phone are leading to a radical reconfiguration of social coordination. Whereas in modernity individuals would coordinate activities in their work and personal lives amongst each other according to schedules, today, Ling observes, we are able to move away from the mediating system of the schedule toward direct contact between individuals. In part, this emerges as transportation systems themselves become more complex and more prone to delays. As recently as two decades ago, being in transit meant being out of touch, but today midcourse adjustments can be made rapidly amongst individuals. If a parent is stuck in traffic, the other one can forego a shopping trip to pick up a child from school. As a consequence of the slack that the mobile phone gives to time-based agreements, Ling notes that schedules have softened and, as it is possible to notify others that one is running late while en route, tardiness is more acceptable.[13]

With all of the comforts available inside the networked car (called variously a Swedish, German, or Japanese phone booth), the automobile easily accommodates the spatial function that Ichiyo Habuchi has deemed the telecocoon. In relation to mobile phone use by Japanese teenagers, Habuchi describes the telecocoon as a virtual networked space created by young friends and lovers out of a constant, steady stream of conversation that keeps them in touch even when they are apart. The telecocoon maintains intimacy at a distance, facilitating private encounters in public spaces. Instead of an architectural plan or spatial design, the telecocoon relies on networking technology to create private space, thereby overcoming the problems that distance introduces into our lives. In Japan, Mizuko Ito observes, the home is too family oriented and too crowded to accommodate friends, so teens resort to their mobile phones, or *keitai,* to text their close friends, maintaining silent conversations the entire time they are away from their friends. *Keitai,* Kenichi Fujimoto writes, are "territory machines" capable of redefining the notion of public space, transforming a subway train seat, a sidewalk, a street corner into the user's "own room and personal paradise."[14] In other countries, however, where talking aloud is more acceptable, Rich Ling observes that "forced eavesdropping" can be an embarrassment to the involuntary audience, the phoner being so absorbed in conversation that they he or she never becomes aware of the context.[15]

Mobile phone use has skyrocketed from 5 million subscribers in the United States (11 million worldwide) in 1990 to 225 million (2.7 billion worldwide) in 2007.[16] Forbes rightly calls it "the most personal and ubiquitous gadget ever devised."[17] Part of daily life in Japan and Europe, the telecocoon has only recently spread widely in the United States as mobile companies have begun providing inter-carrier SMS transmission and promoting the services, most notably by enabling SMS voting through the *American Idol* television program.[18] Mobile companies have also extended the telecocoon to a tele-umbilical for a younger set. When the Firefly was introduced in 2005 as the "mobile phone for mobile kids," it was the first such device designed and marketed for elementary school children (small enough for a child's palm, with flashing lights, a glowing body, and just five keys, easily preprogrammable, with parental and emergency phone numbers). Marketed as a safety device—like placing a tracking sensor in a kid's hand or pocket—the Firefly tethers children to the reach of the parental voice. Perhaps with the Firefly, parents will feel safe letting their children roam the streets alone again.[19]

Epitomized by the iPhone, mobile phones are increasingly capable devices, with digital cameras, games, e-mail and Web access, and video playback capability already integrated into many models. "Text me" may refer interchangeably to a message sent over an instant messaging network such as AOL Instant Messenger (AIM), to a message sent from mobile phone to mobile phone, or even to a message sent from computer to mobile phone. Third generation, or 3G, cellular networks such as EV-DO (Evolution-Data Optimized) and built-in Wi-Fi make it possible for many mobile devices to connect to the Internet at broadband speeds, allowing subscribers to access streaming audio and video and even videoconference. Meanwhile VoIP (Voice Over Internet Protocol) services such as Skype offer alternatives that bypass the mobile carriers' pricey long distance and international rates, an alternative made more viable by handsets that can swap from mobile to Wi-Fi networks.

More recent research by Ito suggests that the widespread introduction of cameras into phones is more likely to have an impact on the way the telecocoon develops as individuals share "an ongoing stream of viewpoint-specific photos with a handful of close friends or an intimate other."[20] It may be that in the future we won't see complex gadgets trying to be all things to all people, but rather more devices like the Sony PSP (PlayStation Portable) gaming platform, a dedicated gaming device capable of connecting to the wireless Internet to network players together or download updates. Although browsers are built into the PSP system to enable connections by games, barring hacking, they remain hidden to users. Instead of one converged device offering one form of access to the network, multiple devices and objects (like cars, toys, and cameras) may acquire network access. Some of these, such as newer Tamagotchi or the Nintendo DS game system, will have their network access limited to ad hoc, local networks, encouraging group use.[21] Similarly, even though the new Apple iPhone readily connects to the Internet over Wi-Fi and runs a version of the Mac OS X operating system, as of this writing it appears likely to be restricted to running only programs the company authorizes.

Shaping their identities through networking technology and living in *keitai*-created telecocoons with their intimate friends, Japanese teenagers are today's *flâneurs.* But American teenagers have followed suit as their social networks have become more and more device-enhanced. The way that Japanese teenagers use their *keitai* is, as Ito points out, contingent on their particular cultural context, but that's precisely the point. Just as the *flâneur* served as a stand-in for broader cultural shifts in modernity, so, too, might the Japanese teenager indicate the symptomatic conditions of early twenty-first-century cultural life, demonstrating how we inhabit localized time and space as well as telematic worlds in which we can be copresent with others at a distance.

Real Virtual Worlds

In his 1831 novel about fifteenth-century France, *Notre Dame de Paris,* Victor Hugo crafted a now famous statement for the lips of the abbot of that church: "This will kill that. The book will kill the building." For Hugo, Gutenberg's marvelous invention put an end to architecture's role as a communication medium. From "the origin of things to the fifteenth century," Hugo wrote, architecture "was the great book of mankind, . . . the principal register of mankind."[22] The printed book, however, was a far more efficient medium for communicating with individuals. In contemporary terms, it had the advantage of broader bandwidth and mobility. Hugo's decision to stage his novel in the fifteenth century was by no means whim: the print literacy that Hugo described beginning its radical spread with Gutenberg only truly became a mass phenomenon in his own day.

If Hugo was largely correct about the capacity of the book to replace the building as text, what about the possibility that the network might replace the building as dwelling place—that virtual space will replace real space?

A decade ago, visionaries such as William Mitchell suggested that with the development of the Internet, the downfall of the modern city was upon us. In their view, the new problem was how to create the "city of bits," the electronically mediated spaces for the lives that we would be leading online, which were as sure to replace the modern city as it, in turn, replaced the village.[23]

The provocative visions of a universal, three-dimensional cyberspace, such as those shown in Michael Benedikt's seminal *Cyberspace: First Steps,* have not come to pass.[24] At the time, with our vision of the future colored by cyberpunk novels like William Gibson's *Neuromancer,* it seemed plausible that we would inhabit virtual cities, our bodies becoming wetware and the spaces and social formations surrounding them increasingly neglected.[25] But VRML (Virtual Reality Modeling Language), the much-touted three-dimensional counterpart to HTML (the hypertext markup language typically used on the Web), hasn't resulted in one commonly used site. Although the Web has become graphically more sophisticated, when we visit it we navigate a two-dimensional interface. Corporate presences on the Internet appear to us as brochureware, not as virtual structures that we can enter into and inhabit. Indeed, the Web is curiously nonspatial, a step back from the use of the desktop and file folders to represent relationships between data.

In retrospect, the all-digital "city of bits" seems to be a historical artifact, the product of a digital culture in which the user was tied to a CRT (cathode-ray tube) screen. The key technological devices that shape our lives—telephones and computers as well as the telematic networks that connect them—are now mobile, free of specific contexts but implicated in situational contexts, coloring those situations just as those situations color their contexts in turn.

Today, however, as the previous sections on place and mobility suggest, rather than having one body withering away in front of the screen, it is progressively more common to navigate two spaces simultaneously, to see digital devices and telephones as extensions of our mobile selves.

But was the prophecy all wrong? To some degree, this world predicted by the techno-futurists has come to pass. The Web is a growing presence in our lives. Mobile phones, e-mail, and browsing for information are increasingly part of the everyday experience of many people. Shopping in particular is becoming more and more virtual for many consumers. If the dot-com crash demonstrated that some business models such as Webvan or Pets.com were not immediately viable, other models have proven more profitable. Bricks-and-mortar stores have been seriously challenged by mail-order megastores such as Amazon.com and content-delivery services such as iTunes that offer lower prices while also making available a wider variety of long-tail products (commodities that are purchased too infrequently to be stocked in normal stores but that collectively rival the share of the market controlled by hits). When we window-shop, it is more frequently in the window of our Web browser.

Moreover, for millions of people, the Internet offers an alternate reality in the form of Massively Multiplayer Online Role Playing Games (MMORPG). Blizzard Entertainment—owners of the most popular MMORPG, World of Warcraft (WoW)—boasts that it has some eight million dedicated players. Even taking into account corporate hyperbole, the fact remains that the population of one game alone exceeds that of Los Angeles. Edward Castronova calls such MMORPGs "the first settlements in the vast, uncharted territory that lies between humans and their machines."[26]

Immersed in these spaces, players occupy avatars, virtual stand-ins for their earthly selves that they can craft to their liking, choosing an appropriate name, hair color and style, clothing, color, gender, race (by this we mean not only in the familiar sense, but elven, dwarven, or orc as well), and pet.

But does the body wither away? As *The Matrix*—the film that, more than any other, is our cinematic allegory for contemporary life—suggests, the flesh and its avatar are linked, and everyday reality and the virtual worlds of the games collide. To take one example, after a player in WoW died of a stroke, her friends organized a virtual funeral for her. But they did so in a contested zone on a WoW server in which players routinely fight other players. A hostile party mounted a raid against the grieving teammates and posted the video on the Net, causing a controversy among gamers. If monsters generated by the game's algorithms had perpetrated the ambush, the players would not have complained, but that other players did so raised questions about moral behavior and the limits of reality in MMORPGs.[27]

Conversely, since MMORPGs are typically based on economic models in which characters generate virtual currency by killing monsters or completing other tasks and spend that currency on virtual items, many millions of dollars a year are generated in the buying and selling of these virtual goods. Even though the legality of the activity has been called into question by some MMORPG companies, this has led to the development of *gold farming,* in which individuals, working for low wages in China and Indonesia, obtain gold or rare virtual items that they then sell for real money. In China some one hundred thousand workers spend twelve-hour days farming gold, but this isn't merely a question of outsourcing—estimates suggest that there are more players of MMORPGs in China than in any other country and the Chinese government estimates that some twenty-four million individuals played online games in 2005.[28]

Even though they are still rather early in their development, MMORPGs seem to have the capacity to feed back into real culture. John Seely Brown and Douglas Thomas suggest that World of Warcraft effectively teaches players how to manage teams in the successful accomplishment of complex tasks.[29] MMORPGs such as Second Life make possible online meetings of individuals dispersed in space and time. Such meetings would be more effective than videoconferencing, it is argued, since even if a participant is replaced by an avatar, the full range of three-dimensional motion in an MMORPG affords a more intimate experience than the flattened world of videoconferencing.

MMORPGs have yet to become mobile. Developers have produced alternate reality games (or ARGs) such as the viral marketing stunt "I Love Bees," but these have so far failed to capture a broad following.[30] Still, as MMORPGs continue to rise in popularity, they suggest another aspect of the quality of network culture: that increasing numbers of us have, or will have, alter egos that dwell as much in virtual, networked worlds as in this one.

The Network and Its Socio-Spatial Consequences

Throughout both this chapter and the book as a whole, we observe how network culture's focus on the node's position in a broader (technological and social) network has supplanted digital culture's drive to abstract the world into discrete, computable elements. The transition toward network culture is not merely technological, it is deeply tied into societal changes. In *The Rise of the Network Society,* Manuel Castells analyzes how society is moving toward more networked forms of organization in production, power, and experience. Corporations, financial markets, criminal activities, and political groups that were structured as vertically integrated hierarchies in modernity are organized as networks in our own time.[31] This condition is by no means placeless. On the contrary, Saskia Sassen has identified the "global city" as the key site for the new global economy. The global city, she concludes, is "a function of a network of cities" that takes precedence over any individual role that these cities might play. In Sassen's analysis, these key metropolitan areas do not function independently but rather act as nodes in a planetary economic system—highly concentrated sites in which interpersonal communications take place and which are intimately connected in a single global economic and communicational network.[32]

The social infrastructure emerging in the global city is augmented by a concentration in network topology. Far from the mythical distributed ideal that ideologists of technology claim it to be, the network has its own physicality, its own material presence. Networks rely on relatively few high-bandwidth transcontinental and transoceanic fiber-optic lines, on even fewer Tier-1 carriers that sell space on these lines, and on still fewer mobile-phone operators and last-mile connection (DSL or cable broadband service) providers that allow the end user to access bandwidth. Interchanges between such networks occur at only a few major peering points, usually one or two major carrier hotels per metropolitan area. This highly centralized system produced by historical factors (most notably the monopoly stature of AT&T prior to divestiture) helps to further concentrate the global city. Not only is this system vulnerable to natural or man-made disasters but, as the scandal over National Security Agency data mining during the summer of 2006 demonstrated, it's all too easy to take advantage of by individuals or governments.[33]

Much as the telecocoon functions on an individual level, the global city's connections create local disconnections. The new space of flows is constituted as a set of hubs and nodes. Areas and populations outside of this logic are subject to the tunnel effect: they virtually don't exist as far as the network and, hence, the dominant world economy is concerned.

Both Castells and Sassen raise concerns about areas that are left out of this network on a planetary, national, and urban scale. Nevertheless, during the last decade, networking technology has had an impact on areas that had previously

Box 1.2 Material and Immaterial Real Estate

From Stephen Graham, "Excavating the Material Geographies of Cybercities," in The Cybercities Reader (London: Routledge, 2004), 139.

> These days, telecommunications and digital media industries endlessly proclaim the "Death of Distance" and the "ubiquity of bandwidth." Paradoxically, however, they actually remain driven by the old-fashioned geographic imperatives of putting physical networks (optic fibres, mobile antennas and the like) in trenches, conduits and emplacements to drive market access. The greatest challenge of the multiplying telecommunications firms in large cities is what is termed the problem of the "last mile": getting satellite installations, optic fibres, and whole networks through the expensive "local loop." In other words, the challenge is to thread networks under the congested roads and pavements of the urban fabric, to the smart buildings, dealer floors, headquarters, media complexes and stock exchanges that are the most lucrative target users. Because high-bandwidth networks have to be end-to-end it is not enough to construct networks to main exchanges and city cores; fibre must be threaded through the curbs of users and beyond the actual computers inside buildings. Consequently, fully 80% of the costs of a network are associated with this traditional, messy

business of getting it into the ground in highly congested, and contested, urban areas.[a] This hard material basis for the "digital revolution" is neglected but crucial. Focusing on it allows analysis to begin to reveal the complex social and technological practices that surround and support the explosion of digitally mediated economic and cultural flows.[b]

Such an approach also begins to reveal the subtle and powerful reconfigurations of urban space that are the result of such changing technological practices. Take an example. In a frenzied process of competition to build or refurbish buildings in the right locations for booming new media, telecommunications, and e-commerce companies, a New York agent reported recently that "if you're on top of an optic fibre line, the property is worth double what it might have been."[c] With this, and many other examples of the reconfigurations of urban space, we see that the "information age," or the "network society," is not some immaterial or anti-geographical stampede on-line. Rather, it encompasses a complex and multifaceted range of restructuring processes that become highly materialized in real places, as efforts are made to equip buildings, institutions, and urban spaces with the kinds of premium electronic and physical connectivity necessary to allow them to assert nodal status within the dynamic flows, and changing divisions of labour, of digital capitalism.[d]

These restructuring processes are intrinsically bound up with changing governance and power relations and patterns of uneven development at all spatial scales, from the transnational to the body.[e] In general, they tend to support a complex fracturing of urban space as premium and privileged financial, media, corporate and telecommunications nodes extend their connectivity to distant elsewheres whilst stronger efforts are made to control or filter their relationships with the streets and metropolitan spaces in which they locate (through defensive urban design, closed circuit surveillance, the privatization of space, intensive security practices, and even road closures).

---

a. See Steve Pile, "The un(known) city . . . or, an Urban Geography of What Lies Beneath the Surface," in *Unknown City: Contesting Architecture and Social Space,* eds. Iain Borden, Jane Rendell, Joe Kerr with Alicia Pivaro, 263–278 (Cambridge, MA: The MIT Press, 2001).

b. Rae Zimmerman, "Social Implications of Infrastructure Network Interactions," *Journal of Urban Technology* 8, no. 3 (December 2001): 97–119.

c. Brant Bernet, "Understanding the Needs of Telecommunications Tenants," *Development Magazine* (Spring 2000): 16–18.

d. Manuel Castells, *The Rise of the Network Society* (Oxford, UK: Blackwell, 1996); Dan Schiller, *Digital Capitalism: Networking the Global Market System* (Cambridge, MA: The MIT Press, 1999).

e. Michael Peter Smith, *Transnational Urbanism* (Oxford, UK: Blackwell, 2001).

been outside the global economy. Developing countries with well-educated, technologically adept, and often English speaking workers such as India, China, Ireland, and Estonia have become hosts for outsourced information work. This provides much needed capital and employment, as well as an infrastructural (physical, human, and organizational) framework that spurs indigenous information economies to emerge. In turn, this has led to tensions in the developed world as jobs are being lost or outsourced. But networking technology has also allowed resistance movements, non-governmental organizations (NGOs), and other bottom-up entities to band together worldwide, creating a powerful antiglobalization movement that seeks to redress the inequalities of network society.[34]

Castells suggests that the network has to be seen as part of a bipolar opposition between "the Net and the self," in which individuals relentlessly try to affirm their identities in a rapidly changing world. This identity formation increasingly happens within networks that are both physical and virtual, filled with individuals who both produce and consume, taking advantage of new kinds of online cultural production.[35]

Online social network services such as Friendster and MySpace tap into this increasingly networked culture. Particularly aimed at young people, social network services are generally not composed of static pages but rather are sites of social interaction that are constantly revisited by their active members. Typically, these sites consist of profile pages that contain photos, demographic information, an individual's personal preferences, a blog or link to a blog, and—in sites operating according to the circle of friends model—links to profiles of an individual's friends as well as comments from friends. Writing about MySpace, danah boyd suggests that such sites are not just pages or media but actual places that take over the site of the teenage hangout. She concludes that, when combined with instant messaging, these sites provide intimate communities that fulfill a vital function for teens who have no real spaces in which to gather.[36]

Always eager to understand and exploit changes in society, marketers have been forced to rethink the way they conceive of their audience to adapt to this new condition. In particular, the new place-based field of geodemographic targeting and profiling combines research into networks, places, and cultural production. Specializing in geodemographic marketing over the last thirty years, the Claritas corporation has mapped the breakdown of the mass audience into some sixty-six distinct demographic clusters based on age, ethnicity, wealth, urbanization, housing style, and family structure. Increasing diversification from immigration, economic changes, and greater choice produce a landscape composed of radically small minorities, the largest a mere three percent of the American population.

Lifestyle differences between clusters can be extreme, and an individual's values and interests depend on their cluster. Members of the upper-middle-class Young Digerati cluster who, for example, inhabit the Fairfax area of Los Angeles might live in a trendy condo, drive a Toyota Hybrid, and generally vote liberal while maintaining a libertarian bent. This could be a family with young, tech-savvy parents who work in digital production for Hollywood and are drawn to the area by the array of coffeehouses. But only forty-five miles away, in the postsuburban Orange County town of Newport Beach, another couple from the same socioeconomic background (both couples are likely to be Caucasian, Asian, or mixed, and have finished graduate school) and same salary might be members of the Blue Blood Estates cluster—business executives who value their newly built McMansion homes, vote Republican, enjoy golf vacations, and eat fast-food that their children prefer.

These new tribes are widely dispersed nationally and globally, connected to other members of their own tribes by telecommunications and media. Still, on a local scale, people live next to people they like. Clusters exist as small geographic communities. Although clusters always overlap one another—Claritas generally identifies five clusters per area—they do so in locations that suit them infrastructurally. Individualist, extremely liberal, often gay, arty singles will seek urban communities with a vibrant street life—places like West Hollywood and Silver Lake in the Los Angeles area, Lakeview in Chicago, or Dupont Circle in Washington DC—not manicured homes in postsuburbia. Place and community are themselves forms of infrastructure today, key devices in the network.[37]

Steven Johnson suggests that the renewed interest in cities during the 1980s and 1990s will only increase with the growth of what Chris Anderson calls the *long tail.* Anderson observes that the demand curve for cultural products has traditionally been understood as validating the production of a small number of hits to be bought up by a vast consumer market. In his theory of the long tail, Anderson suggests that the Internet is making the flat part of the long tail—populated by products appealing to ever-smaller niches—as profitable as the head. According to Anderson, tools such as aggregators and search engines couple with a societal shift in media consumption to the flat part of the long tail to increasingly leave behind a one-size-fits-all mentality for an interest in more eccentric, niche tastes. Johnson argues that with culture moving to the flat part of the long tail, the diversity of taste cultures that we can find in dense cities will appeal to us more and more.[38]

Geospatial Web and Locative Media

Place itself does not disappear in favor of the "city of bits." On the contrary, place is as important as ever, playing a key role in the network itself. Still, the previous examples of recent changes in our relationship to the spaces that surround us are all dominated by the seemingly inescapable logic that the price of new connections is local disconnection. But this logic may soon be changing. Two emerging technologies—the geospatial Web and ubiquitous computing—suggest an intertwining of the network and the local, bringing with it new possibilities and new questions.

Geographic Information Systems (GIS) offer a way to represent and analyze data spatially and have been commonly used by industry and government for some time now. Developed in 1967 by Roger Tomlinson at the Canadian Department of Energy, Mines and Resources, the first such system, called CGIS (for Canadian GIS) collated information on land use.[39] Since then, government agencies such as the U.S. Geological Survey and the U.S. Census Bureau, as well as corporations utilizing geodemographic marketing, have adopted GIS with enthusiasm. Much as the spreadsheet revolutionized businesses by making it possible to test scenarios on a personal computer, GIS make it possible to model and hypothesize geospatial scenarios such as changes in a watershed due to construction, the spread of a plume of fuels and solvents underneath an airport and the surrounding neighborhood, the rise or fall of a city's tax base as a result of a new park, or shifts in congressional seats caused by redistricting. For forecasting and analyzing this kind of information, GIS is now indispensable. For the most part, however, the specialized nature of GIS has largely meant that the administration, development, and use of such data has been the province of government, corporations, NGOs, and other research organizations.

Over the last few years, commonly used Internet tools have made GIS available to end users, offering what Institute of the Future researcher Mike Liebhold has called the *geospatial Web*—an augmentation of the placeless information of the browser-based Internet with geographic coordinates.[40] Internet mapping sites such as MapQuest, Yahoo! Maps, and Google Maps are the most familiar applications of GIS technologies, offering user-definable maps and door-to-door, turn-by-turn driving directions. To be fair, turn-by-turn maps, such as the American Automobile Association's Triptiks, have been available for almost a century. If the online services make such information more convenient and more readily accessible by dispensing with the map in favor of point-to-point travel, they also enhance the tunnel effect of networks. With a map, one might be tempted to go off route to see a nearby attraction, but with turn-by-turn directions, routes are optimized and only the most prosaic sponsored businesses interrupt the smooth flow of one's drive.[41]

In the case of Google Maps, Google has made it possible for amateur programmers to interface the site's data and maps easily. As a result, programmers have created hundreds, if not thousands, of Google Maps mash-ups—geospatial interfaces to all manner of information interesting to end users such as free Wi-Fi nodes (http://www.gwifi.net/), real estate available on Craigslist (http://www.housingmaps.com/), locations of cellular towers (http://www.cellreception.com/towers/index.html), or airports in which pets have been lost, injured, or killed (http://www.petflight.com/incidents/map).

Google is also responsible for the Google Earth application, dubbed "the People's GIS" for its attractive, easy-to-use interface capable of rendering three-dimensional flyovers based on satellite photographs and contour data in real time; for its ability to display layers of GIS data—such as locations of shopping malls, monuments, places to eat or sleep, or city boundaries; and for the ease by which users input their own information such as coordinates to Wikipedia articles, earthquakes, or annotations to historic or interesting sites (e.g., airplanes visible in the satellite photographs, crime scenes, or corporate signs).[42] To some degree, Google Earth gives a taste of a future *digital earth,* a term coined by Vice President Al Gore in 1998 to refer to a three-dimensional virtual representation of the planet that would allow individuals to explore scientific and cultural information about the planet.[43] But even though Google Earth is a fascinating application, it does not have the depth to allow users to find out the natural or human history of a site. Lacking any real purpose, Google Earth has had little impact on everyday life in comparison to the more prosaic two-dimensional mapping interface of Google Maps and may yet replay the failure of three-dimensional Web interfaces.

The Holy Grail for networked place, however, is to take GIS information mobile. With GPS technology improving and both mobile phones and personal digital assistants (PDAs) gaining Internet connectivity, hackers, software developers, and artists alike have sought to turn the model of non-place on its head by using networking technology to create social connections. This locative media is based on the promise of handheld location-aware devices that can interface with the geospatial Web to provided georeferenced information on the spot to end users. Proponents hope that inclusion of geographic references on the Web, and the delivery of that data to the mobile end user, will make it possible for digital media to be associated with a site, or literally found there. Thus, comments, blog entries, restaurant reviews, past photographs, real estate prices, and such would be available at the sites they are associated with. This

Box 1.3 Locative Media and the Threat of Tracking

From Jordan Crandall, "Operational Media," Arthur and Marilouise Kroker, eds., CTHEORY, http://www.ctheory.net/articles.aspx?id=441.

In media-saturated societies, surveillance has gradually been made "friendly" and transformed into spectacle, to the extent that it is no longer a condition to be feared. Rather, it is a condition to be courted: witness the phenomena of reality television, blogs, and webcams, and the rise of the media mise-en-scene as the primary form of social authentication.[a] In recent cyber discourses, this "friendly" control is often regarded as self-regulating: we are an integral part of systems that self-adjust through market dynamics or adaptive behaviors, allowing for the emergence of new forms of maneuver and masquerade. Within new ecologies of mind,[b] we benefit from machine-human interactions all around us, a pervasive web of shared resources that offers boundless opportunity for identity refashioning. Further: in a database-driven culture of accounting, one needs to appear on the matrices of registration in order to "count." To be accounted for is to exist.

Perhaps nowhere have the contradictions of communicative opportunism/surveillant precision been made more palpable than in new portable wireless devices, especially those that are increasingly "location-aware." These technologies, along with their semiotics and uses, are serving to weave together degrees of temporal and spatial specificity, against the grain of much of the "delocalized" orientation of virtual discourses during the last decade—but perhaps more true to the strategic origins of the cybernetic tradition, which was, after all, concerned with the precise calculation of position. . . .

The potential of GPS-enabled devices, ubiquitous transponders, and other locationing technologies present a world where every object and human is tagged with information specifications including history and position—a world of information overlays that is no longer virtual but wedded to objects, places, and positions, and no longer fully simulative since it facilitates an active trafficking between model and reality. Such location-specific technology combines information, movement, and precise positioning—knowing "where" as well as "what."

These technologies and their discourses aim to increase productivity, agility, and awareness, yet they vastly increase the tracking capabilities of marketing and management regimes. You are able to get what you want faster, but your behavior is tracked and analyzed by marketers who also can provide this information to police and military sources, who increasingly depend upon the business sector for a large part of their intelligence. (After the carnage of the Civil War, the U.S. military was prohibited from future interventions into the domestic realm. Since most of the spy satellites are owned by the military, the military "outsources" some of its domestic intelligence needs to commercial satellite providers, while relying on data gathered through the private sector on a number of fronts, especially to meet the sudden growth in intelligence demands after 9/11.) Information from buying habits, travel locations, and audience demographics can be integrated into one comprehensive system, which aims to target consumers at the one-to-one level, offering individually-tailored enticements. Tracked, the user becomes a target within the operational interfaces of the marketing worlds, into whose technologies state surveillance is outsourced.

---

a. See Peter Weibel, "Pleasure and the Panoptic Principle," and Ursula Frohne, "Screen Tests: Media Narcissism, Theatricality, and the Internalized Observer" in *{CTRL}SPACE: Rhetorics of Surveillance from Bentham to Big Brother,* eds., Thomas Levin, Ursula Frohne, and Peter Weibel, 215–219; 253–77 (Cambridge, MA: The MIT Press, 2002).

b. See Gregory Bateson, *Steps to an Ecology of Mind* (Chicago: University of Chicago Press, 2000), 466.

is already possible with the Vindigo service, which provides locations, contact information, maps, and thumbnail reviews of restaurants, bars, bathrooms, services, museums, galleries, music venues, and so on for major metropolitan areas in the United States, as well as London, to PDA owners and mobile phone users.[44] Other locative media services propose location awareness for social networking. At Dodgeball, users sign up their mobile phone numbers with the service and inform their friends that they are doing so. Later, to connect with friends, the user can "check in" by notifing the service of her or his location with an SMS text message. The service then sends this information to mobile phones of the user's friends, as well as friends of friends, that are within a ten-block radius.[45]

Melding the geospatial Web with locative media promises that you can leave your mark on the world or read the marks others leave behind, re-creating place in a Borgesian digital map. Artifacts and places will be imbued with memories in a far richer way than ever before. Given a geocoded, Wikipedia-like interface, it is possible to imagine the entire world annotated with histories, becoming as Freud once wrote of the mind in *Civilization and Its Discontents,* a place "in which nothing once constructed had perished, and all the earlier stages of development had survived alongside the latest."[46]

But what of forgetting in this age of locative media? Will this lead to an accumulation of mindless geospatial data-junk that buries spaces? Some personal memories—such as traumatic events—might be better left forgotten. Moreover, locative media have failed to win many adherents and remain in a perpetually about-to-happen future. For its part, Vindigo has not developed any new features in years, and Dodgeball seems too convoluted for the average person to use. On a recent visit, the site advertised that the top user in New York City had checked in merely seventeen times in the last ten days.

RFIDs, Ubiquitous Computing, and the Coming Sentience of the World

If locative media offers sentient users the ability to augment place, other developments—some of which are already in widespread practice—suggest that the less-than-sentient world may soon gain a degree of awareness. Already widely in use, RFID tags are a passive way of giving objects—but also people—the capacity to tell their stories. RFIDs are small tags, sometimes cunningly disguised, that require no internal power source but respond to radiofrequency queries from transponders with a distinct signal and are commonly used for inventory tracking in stores. As each RFID has a unique identifier, it can forever be associated with a distinct object. It's a small leap to imagine that RFIDs could also be tagged by their owners so that their stories can be added to. In his book *Shaping Things,* Bruce Sterling suggests that RFIDs could have a positive use in creating spimes, his neologism for objects that can be tagged with cradle-to-grave information about where they have been, where they are, and where they are going. The origin, conditions of manufacture, and ultimate destination of an object can all potentially be tracked through its RFID.[47]

The result of this utopian vision is to make visible a genealogy of objects for ecological and political purposes. As Walter Benjamin once wrote:

> The cultural heritage we survey has an origin that we cannot contemplate without horror: it owes its existence not merely to the effort of great geniuses who created it, but to the anonymous toil of their contemporaries. There is not a single artifact of culture that is not simultaneously an artifact of barbarism. And just as no artifact is free of barbarism, so too the process of its reception, by means of which it has been passed on from one recipient to the next, is equally fettered.

Being aware of such a Benjaminian genealogy of the object, Sterling suggests, might lead us to radically rethink the social and ecological impact of our purchases.[48]

But RFIDs have a dark side. In *Spychips,* Katherine Albrecht and Liz McIntyre suggest that they are the gravest of threats to privacy. Already in 2001, they observe, IBM had a patent for tracking individuals with RFIDs, and Verichip has developed and received approval from the FDA for human-implantable RFIDs. Chillingly, Verichip suggests that implanted RFIDs could be used to track guest workers in the United States, conjuring a future out of *Logan's Run.*[49] But Albrecht and McIntyre observe that RFIDs don't need to be implanted to track us. Since RFIDs are used as theft-prevention devices, they are, like some virulent form of insectile parasite, hard to destroy and generally invisible, lurking in the products you wear or bring wherever you go, ready to give themselves up to a radio-frequency query and, if purchased with your credit card, forever identifiable with you. Assuming a will on the part of marketers or the government, it is trivial to construct a *Minority Report* style system that would use RFID-bearing clothing and personal items to actively track you through your daily travels. As yet, Albrecht and McIntyre observe, no foolproof way for deactivating or killing undetected RFIDs has been identified.[50]

Tracking individuals using RFIDs is already possible. At LEGOLAND® Billund, the KidSpotter service, introduced in 2004 by the amusement park and the Tryg insurance company, employs RFIDs to help parents keep track of their children. Children enrolled in the system wear a special Kidspotter wristband with a tag the size of a matchbook attached. When parents send an SMS message to the system, they receive a return message containing their children's coordinates, which they can then check against a special map of the park.[51]

But RFIDs still suggest that objects will be passive. The RFID is a passive tag waiting to be activated. Things, however, are beginning to acquire a degree of contingency, gaining the ability to talk back. In the 2005 report *The Internet of Things,* the International Telecommunications Union predicts a "new era of ubiquity" available anywhere, anytime that will permit connections between humans and things (H2T) and between things themselves. Ultimately, thing-to-thing (T2T) communication will circumvent communicative networks between humans.[52]

In this scenario, the formerly inanimate other will be able to report back about its location, condition, and needs. Things will cease to be mere objects, commodities, or fetishes valued by humans. Exceeding anything that Karl Marx could have ever imagined, things will become active, even sentient, observers, able to communicate with each other as much as with us. Our age-old animist dreams of a world imbued with spirits and personalities may be around the corner.

Like the animist spirits of the *genius loci,* locative media, RFIDs, and the Internet of things are premised on their invisibility, on a near future in which an invisible data overlay blankets the earth. Meanwhile, not only devices (like mobile phones and PDAs), but also furniture (like chairs and tables), objects (like trees and street signs), buildings (like monuments and apartment buildings), and landscapes (like forests, deserts, and riverbeds) become sentient information platforms—sensors to collect and send data to whoever is out there to collect, analyze, and read it.

And like all spirits, these could be dangerous; not only do we have the dystopian scenarios of the geospatial Web filled with geo-spam and our every

Box 1.4 Turning off Ubiquitous Computing

From Adam Greenfield, "Thesis 77 Everyware must be deniable," in Everyware (Berkeley: New Riders, 2005), 246–247.

Our last principle is perhaps the hardest to observe: ubiquitous systems must offer users the ability to opt out, always and at any point.

You should have the ability to simply say "no," in other words. You should be able to shut down the ubiquitous systems you own and face no penalty other than being unable to take advantage of whatever benefits they offered in the first place. This means, of course, that realistic alternatives must exist.

If you still want to use an "old-fashioned" key to get into your house, and not have to have an RFID tag subcutaneously implanted in the fleshy part of your hand, well, you should be able to do that. Should you want to pay cash for your purchases rather than tapping and going, you should be able to do that too. And if you want to stop your networked bathtub or running shoe or car in the middle of executing some sequence, so that you may take over control, there should be nothing to stand in your way.

In fact—and here is the deepest of all of the challenges these principles impose on developers and on societies—where the private sphere is concerned, you should be able to go about all the business of an adult life without ever once being compelled to engage the tendrils of some ubiquitous informatic system.

In public, where matters are obviously more complicated, you must at least be afforded the opportunity to avoid such tendrils. The mode of circumvention you're offered doesn't necessarily have to be pretty, but you should always be able to opt out, do so without incurring undue inconvenience, and above all without bringing suspicion onto yourself. At the absolute minimum, ubiquitous systems with surveillant capacity must announce themselves as such, from safely beyond their field of operation, in such a way that you can effectively evade them.

The measure used to alert you needn't be anything more elaborate than the signs we already see in ATM lobbies, or anywhere else surveillance cameras are deployed, warning us that our image is about to be captured—but such measures must exist.

Better still is when the measures allowing us to choose alternative courses of action are themselves networked, persistently and remotely available. Media Lab researcher Tad Hirsch's Critical Cartography project is an excellent prototype of the kind of thing that will be required: it's a Web-based map of surveillance cameras in Manhattan, allowing those of us who would rather not be caught on video to plan journeys through the city that avoid their field of vision. (Hirsch's project also observes a few important provisions of our principle of self-disclosure: his application includes information about where cameras are pointed and who owns them.)

All of the wonderful things our ubiquitous technology will do for us—and here I'm not being sarcastic; I really do believe that some significant benefits await only our adoption of this technology to appear—will mean little if we don't, as individuals, have genuine power to evaluate its merits on our own terms and make decisions accordingly. We must see that everyware serves us, and if and when it does not, we must be afforded the ability to shut it down. Even in the unlikely event that every detail of its implementation is handled perfectly and in a manner consistent with our highest ambitions, a paradise without choice is no paradise at all.

move tracked with RFIDs, but what might happen if every light bulb insisted on leaving behind its life story or if a printer reported on what you printed?[53]

Conclusion

Today, Augé's solitary non-places are an artifact of the past. We will never be alone again, except by choice. It is likely, however, that new forms of disconnect and alienation will arise. Being too connected may become more of a problem for us than loneliness. Dwelling in the virtual—be it in the World of Warcraft or on a Blackberry—can already be a dangerous addiction that destroys families.

Global connections versus local disconnections, the growing overlap of local and virtual presences, telecocooning, the emergence of real virtual worlds, and the suggestion that locative media will utterly reconfigure our relationship with place all offer opportunities as well as challenges. Place, it seems, is far from a source of stability in our lives, but rather, once again, is in a process of a deep and contested transformation.

Notes

1. The public sphere is "made up of private people gathered together as a public and articulating the needs of society with the state." Habermas, *The Structural Transformation of the Public Sphere: An Inquiry into a Category of Bourgeois Society* (Cambridge, MA: MIT Press, 1991), 176.

2. Baudelaire, "The Painter of Modern Life," in *My Heart Laid Bare and Other Prose Writings,* 34 (London: Soho Book Company, 1896).

3. See Benjamin, "The Paris of the Second Empire in Baudelaire" in *Charles Baudelaire: A Lyric Poet in the Era of High Capitalism.* (London: New Left Books, 1977), 37; Anne Friedberg, *Window Shopping: Cinema and the Postmodern* (Berkeley: University of California Press, 1994), 36–37.

4. Simmel, *Soziologie: Untersuchungen Über Die Formen Der Vergesellschaftung* (Berlin: Duncker & Humblot, 1958), 486.

5. See Richard Sennett, *The Fall of Public Man* (New York: Alfred A. Knopf, 1974); Guy Debord, *The Society of the Spectacle* (New York: Zone Books, 1994).

6. Jacobs, *The Death and Life of Great American Cities,* (New York: Vintage, 1961). See also Lynn Spigel, *Make Room for TV: Television and the Family Ideal in Postwar America* (Chicago: University of Chicago Press, 1992); Spigel, *Welcome to the Dreamhouse* (Durham, NC: Duke University Press, 2001).

7. Augé, *Non-Places: An Introduction to the Anthropology of Supermodernity* (New York: Verso, 1995).

8. Brian Niemetz, "Café Regulars Work With Perks," *New York Post,* May 21, 2007, http://www.nypost.com/seven/05212007/atwork/coffee___to_stay_atwork_brian_niemietz.htm.

9. Ithiel de Sola Pool, *The Social Impact of the Telephone* (Cambridge, MA: MIT Press, 1977).

10. Scannell, *Radio, Television and Modern Life: A Phenomenological Approach* (Oxford, UK: Blackwell Press, 1996), 76.

11. Abler, "What Makes Cities Important," *Bell Telephone Magazine* 49, no. 2 (1970): 10–15.

12. Baudrillard, "The Ecstasy of Communication," in *The Anti-Aesthetic,* ed. Hal Foster, 127 (Seattle: The Bay Press, 1985).

13. Ling, *The Mobile Connection: The Cell Phone's Impact on Society* (San Francisco: Morgan Kaufmann, 2004), 57–82.

14. Habuchi, "Accelerating Reflexivity," in *Personal, Portable, Pedestrian: Mobile Phones in Japanese Life,* eds. Mimi Ito, Daisuke Okabe, and Misa Matsuda, 179 (Cambridge, MA: MIT Press, 2005); Kenichi Fujimoto, "The Third-Stage Paradigm: Territory Machines from the Girls' Pager Revolution to Mobile Aesthetics," ibid, 10.

15. Ling, *The Mobile Connection,* 123–143.

16. eTForecasts, "Cellular Subscriber Forecasts by Country," http://www.etforecasts.com/products/ES_cellular.htm.

17. Erika Brown, "Coming Soon to a Tiny Screen Near You," *Forbes.com,* http://www.forbes.com/forbes/2005/0523/064.html.

18. Brad Smith, "Texting Knows No Bounds," *Wireless Week,* October 15, 2005, http://www.wirelessweek.com/texting-knows-no-bounds.aspx.

19. Contrast with David Derbyshire, "How Children Lost the Right to Roam in Four Generations," June 15, 2007, *The Daily Mail,* http://www.dailymail.co.uk/pages/live/articles/news/news.html?in_article_id=462091.

20. Ito, "Intimate Visual Co-Presence," (paper, UbiComp 2005, International Conference on Ubiquitous Computing, Tokyo, Japan), http://www.itofisher.com/mito/archives/ito.ubicomp05.pdf.

21. See Ito, "My First DS Backchannel," Mizuko Ito Weblog, http://www.itofisher.com/mito/weblog/2006/05/my_first_ds_backchannel.html; Ito, "Gotchi Networks,"

Networked Publics Weblog, http://networkedpublics.org/portable_media/gotchi_networks.

22. Hugo, *Notre-Dame de Paris* (New York: Penguin Books, 1978), 188 and 195.

23. William Mitchell, *City of Bits: Space, Place, and the Infobahn,* (Cambridge, MA: MIT Press, 1995).

24. Benedikt, *Cyberspace: First Steps* (Cambridge, MA: MIT Press, 1991).

25. Gibson, *Neuromancer* (New York: Ace, 1984).

26. Castronova, *Synthetic Worlds: The Business and Culture of Online Games* (Chicago: University of Chicago Press, 2005).

27. Joe Rybicki, "The Real and the Semi-Real," Joe Rybicki's 1UP Blog, http://www.1up.com/do/blogEntry?bId=6883235&publicUserId=4553267; "Rival Guild crashes WoW funeral (video!)," NeoGAF forums, http://www.neogaf.com/forum/showthread.php?t=94595.

28. See James Lee, "From Sweatshops to Stateside Corporations, Some People are Profiting off of MMO Gold," 1UP.com, http://www.1up.com/do/feature?cId=3141815; Paul, "Secrets of Massively Successful Multiplayer Farming," http://www.gameguidesonline.com/guides/articles/ggoarticleoctober05_01.asp; David Barbazoa, "Ogre to Slay? Outsource it to the Chinese," *The New York Times,* December 9, 2005, http://www.nytimes.com/2005/12/09/technology/09gaming.html?ex=1291784400&en=a723d0f8592dff2e&ei=5090&partner=rssuserland&emc=rss.

29. Brown and Thomas, "You Play World of Warcraft? You're Hired!" *Wired* 14.4 (April 2006), http://www.wired.com/wired/archive/14.04/learn.html.

30. On ARGs, see Alternate Reality Gaming Network, http://www.argn.com/.

31. Castells, *The Rise of the Network Society* (Cambridge, MA: Blackwell, 1996).

32. Sassen, "On the 21st Century City," interview by Blake Harris, *Government Technology Interview,* June 1997, http://www.govtech.net/magazine/story.php?id=95352&issue=6:1997.

33. See Kazys Varnelis, "The Centripetal City: Telecommunications, the Internet, and The Shaping of the Modern Urban Environment," *Cabinet Magazine* 17, (Spring 2004/2005): 27–33.

34. Saskia Sassen, *Globalization and its Discontents* (New York: The New Press, 1998).

35. Castells, *The Rise of the Network Society,* second edition (New York: Blackwell Publishers, 2000), 22.

36. danah boyd, "Identity Production in a Networked Culture: Why Youth Heart MySpace," (paper, American Association for the Advancement of Science, February 19, 2006), http://www.danah.org/papers/AAAS2006.html.

37. On Claritas, see Michael J. Weiss, *The Clustered World: How We Live, What We Buy, and What it All Means About Who We Are* (New York: Little, Brown, and Company, 1999).

38. See Anderson, "The Long Tail," *Wired* 12.10 (October 2004), http://www.wired.com/wired/archive/12.10/tail.html; Johnson, "Emerging Technology: Friends 2005: Hooking Up," *Discover* 26, no 9, http://www.discover.com/issues/sep-05/departments/emerging-technology/.

39. Wikipedia contributors, "Geographic information system," Wikipedia, The Free Encyclopedia, http://en.wikipedia.org/w/index.php?title=Geographic_information_system&oldid=100249147.

40. Liebhold, "The Geospatial Web: A Call to Action," *O'Reilly Network,* May 10, 2005, http://www.oreillynet.com/pub/a/network/2005/05/10/geospatialweb.html.

41. See Nick Paumgarten, "Getting There: The Science of Driving Directions," *The New Yorker,* April 24, 2006, http://www.newyorker.com/fact/content/articles/060424fa_fact.

42. "Google Earth: The People GIS," *AECNews.com,* July 11, 2005, http://aecnews.com/articles/1050.aspx.

43. Al Gore, "The Digital Earth: Understanding our Planet in the 21st Century," (speech, California Science Center, Los Angeles, January 31, 1998)1.rtf, http://www.isde5.org/al_gore_speech.htm.

44. For a survey of locative media see Marc Tuters and Kazys Varnelis, "Beyond Locative Media," *Leonardo* 39, no. 4 (August 2006): 357–363.

45. See Johnson, "Emerging Technology: Friends 2005."

46. Freud, *Civilization and Its Discontents* (New York: J. Cape & H. Smith, 1930), 17.

47. Sterling, *Shaping Things,* (Cambridge, MA: MIT Press, 2005).

48. Although it is not linked to RFIDs, Natalie Jerimijenko's "How Stuff is Made" (http://www.howstuffismade.org/) is a wiki-driven visual encyclopedia that aspires to produce similar genealogies and can serve as something of a model for the spime.

49. Tiki Barber, Brian Kilmeade, and Kiran Chetry, "Fox & Friends" (interview with Scott Silverman), Fox News Channel, May 16, 2006, transcript at http://www.spychips.com/press-releases/silverman-foxnews.html.

50. Albrecht and McIntyre, *Spychips: How Major Corporations and Government Plan to Track Your Every Move With RFID* (Nashville: Nelson Current: 1998).

51. Will Sturgeon, "Protecting Your ID: RFID chips on kids makes Legoland safer," *Silicon.com,* June 24, 2004, http://www.silicon.com/research/specialreports/protectingid/0,3800002220,39121670,00.htm.

52. International Telecommunication Union, *ITU Internet Reports 2005: The Internet of Things* (Geneva: International Telecommunication Union, 2005).

53. Printers already have forensic information built into them that is printed invisibly on every page. See "DocuColor Tracking Dot Decoding Guide," Electronic Frontier Foundation, http://www.eff.org/Privacy/printers/docucolor/.

2

Culture: Media Convergence and Networked Participation

Adrienne Russell, Mizuko Ito,
Todd Richmond, and Marc Tuters

The convergence between old and new media is tied to broad-based changes in how power and information are distributed across society, geography, and technology. Combined with low-cost digital authoring tools, pervasive digital networks have lowered the threshold for producing, publishing, and disseminating knowledge and culture. As a result, the boundaries between producer and consumer, and between public and private, are blurring. Through the Internet, casual communication, personal stories and opinion, homegrown news, and amateur cultural works can be made easily available to large audiences. Artifacts once associated with personal culture (like home movies, snapshots, diaries, and scrapbooks) have now entered the arena of public culture (like newspapers, commercial music, cinema, and television).[1] As a consequence, the top-down, one-to-many relationship between mass media and consumers is being replaced, or at least supplemented, by many-to-many and peer-to-peer relationships. New cultural forms employing remix appropriate existing commercial media, raising questions about copyright and control. As individuals rise in influence through news blogs and other emerging forms of communication and old media lose influence, questions of who has authority come to the fore.

In this chapter, we identify four domains that have flourished with the turn toward networked public culture—amateur and non-market production, networked collectivities for producing and sharing culture, niche and special-interest groups, and aesthetics of parody, remix, and appropriation—and trace them through four case studies—news blogs, viral marketing, anime fandoms, and amateur and remix music.

The growth of professional and commercial media in the nineteenth and twentieth centuries ghettoized amateur cultural production. Domains such as amateur musical performance, personal correspondence, diaries, local newsletters, and everyday talk have always been key dimensions of cultural life but were long overshadowed by commercial and professional cultural forms. Mass commercial media created a translocal, star-studded, and spectacular arena of shared public culture and imagination that transcended local cultural forms. But even then, individuals were not just consumers but producers, generating original cultural artifacts and reshaping mass media productions into forms and narratives that fit their own desires.[2] With the advent of networked public culture, the amateur, local, and niche cultures that persisted in the shadow of mass media are gaining greater visibility and translocal reach. Blogs, online chat, auction sites, Web forums, P2P file sharing, streaming online video, and social network services are platforms for longstanding amateur cultural production, fandoms, and everyday communication to flourish in new ways.

Today, as Yochai Benkler theorizes, we are at the beginning of a shift away from commercial media and centrally organized knowledge production toward non-market and distributed production.[3] Amateurs remixing and distributing music over the Internet, fans producing derivative works of fiction and art, marketers appropriating the idioms of viral amateur culture, and bloggers debating the latest news—these are all examples of, in the words of John Hagel and John Seely Brown, "the edge becoming the core."[4] In doing so, they create cultures that both draw from and threaten commercial media. Geert Lovink suggests that these networked publics follow a nihilist impulse against moral absolutes and objective truths, which in media terms translates into a growing distrust for commercial news organizations and their product. He writes, "Questioning the message is no longer a subversive act of engaged citizens but the a priori attitude, even before the TV or PC has been switched on."[5]

Unlike commercial cultural production, which relies on professionalized, institutionalized, and capitalized systems, amateur and non-market production often utilizes more disorganized and socially distributed mechanisms for creating knowledge and culture. Benkler writes of how the processing power of many personal computers can be a source of considerable power when coordinated through smart networks. He considers the example of SETI@home, a scientific experiment that uses Internet-connected computers of worldwide participants in the search for extraterrestrial intelligence, as a model for this kind of distributed processing and knowledge production.[6] Henry Jenkins describes a more socially distributed intelligence in the activities of spoiler groups for the reality TV show *Survivor*. By gathering information from all over the world and communicating over the Internet, networked fan groups

Box 2.1 Nonmarket Cultural Production

From Yochai Benkler, *The Wealth of Networks* (New Haven: Yale University Press, 2006), 6–7.

Human beings are, and always have been, diversely motivated beings. We act instrumentally, but also noninstrumentally. We act for material gain, but also for psychological well-being and gratification, and for social connectedness. There is nothing new or earth-shattering about this, except perhaps to some economists. In the industrial economy in general, and the industrial information economy as well, most opportunities to make things that were valuable and important to many people were constrained by the physical capital requirements of making them. From the steam engine to the assembly line, from the double-rotary printing press to the communications satellite, the capital constraints on action were such that simply wanting to do something was rarely a sufficient condition to enable one to do it. Financing the necessary physical capital, in turn, oriented the necessarily capital-intensive projects toward a production and organizational strategy that could justify the investments. In market economies, that meant orienting toward market production. In state-run economies, that meant orienting production toward the goals of the state bureaucracy. In either case, the practical individual freedom to cooperate with others in making things of value was limited by the extent of the capital requirements of production.

In the networked information economy, the physical capital required for production is broadly distributed throughout society. Personal computers and network connections are ubiquitous. This does not mean that they cannot be used for markets, or that individuals cease to seek market opportunities. It does mean, however, that whenever someone, somewhere, among the billion connected human beings, and ultimately among all those who will be connected, wants to make something that requires human creativity, a computer, and a network connection, he or she can do so—alone, or in cooperation with others. He or she already has the capital capacity necessary to do so; if not alone, then at least in cooperation with other individuals acting for complementary reasons. The result is that a good deal more that human beings value can now be done by individuals, who interact with each other socially, as human beings and as social beings, rather than as market actors through the price system. Sometimes, under conditions I specify in some detail, these nonmarket collaborations can be better at motivating effort and can allow creative people to work on information projects more efficiently than would traditional market mechanisms and corporations. The result is a flourishing nonmarket sector of information, knowledge, and cultural production, based in the networked environment, and applied to anything that the many individuals connected to it can imagine. Its outputs, in turn, are not treated as exclusive property; they are instead subject to an increasingly robust ethic of open sharing, open for all others to build on, extend, and make their own.

collectively produce knowledge that far exceeds what local fan groups could muster.[7] Similarly amateur subtitling groups for Japanese television series rely on globally networked teams to produce and distribute their work. Blogs are another kind of collective intelligence in which individuals pool their fact-finding capabilities to gather knowledge that can challenge the authority of the professional press.

Chris Anderson describes how networked distributors like Amazon.com increasingly make profits not from the short head—a small number of best-sellers—but from the long tail—a wide variety of niche products, each of which has relatively small circulation.[8] Combined with the ability of digital communication to directly connect special-interest groups, these new distribution channels have enabled small producers and small audiences to find one another. The intimate dynamics of local communities can now extend to transnational interest networks. But unlike local communities, which are centered on place-based affiliation, contemporary networks support associations based on esoteric knowledge communities and niche cultural affiliations. The case of anime fandoms is particularly pertinent in this respect. With the Internet, overseas audiences for Japanese animation have exploded in quantity and diversity, aided by distribution from commercial sites like Amazon.com, Netflix, RentAnime, as well as various P2P alternatives. Similar dynamics are at work in the rise of micro-fandoms for alternative and amateur music as well as within the blogosphere.

In addition to the changes in the structures and networks of cultural production, networked public culture has also been associated with particular genres and styles. In this new-media ecology, works that can be produced quickly, at low cost, and that appropriate the products of commercial culture have a new kind of cultural salience. Informal banter about the latest news, music, or television show is published on blogs. Amateur music and video production comment upon existing cultural work. Often employing parody and remix, new forms of cultural production freely combine informal and amateur domains of culture with formal and professional ones. The products of mainstream and commercial culture still retain a certain pride of place in the global imagination; they are the polished and mass-distributed commercial forms that make up what fans call "the canon" or what critics call "corporate media." By mashing up, remixing, playing out alternative narratives, and providing snarky commentary on commercial culture, niche publics can create new cultural forms that draw from both local and translocal referents. Viral political mash-up videos, remixed music, anime music videos, and much of the blogosphere exemplify these aesthetics and discursive styles.

Box 2.2 The Long Tail

From Chris Anderson, "The Long Tail in a Nutshell," Long Tail blog, http://www.thelongtail.com/about.html.

The theory of the Long Tail is that our culture and economy is increasingly shifting away from a focus on a relatively small number of "hits" (mainstream products and markets) at the head of the demand curve and toward a huge number of niches in the tail. As the costs of production and distribution fall, especially online, there is now less need to lump products and consumers into one-size-fits-all containers. In an era without the constraints of physical shelf space and other bottlenecks of distribution, narrowly-targeted goods and services can be as economically attractive as mainstream fare.

One example of this is the theory's prediction that demand for products not available in traditional bricks-and-mortar stores is potentially as big as for those that are. But the same is true for video not available on broadcast TV on any given day, and songs not played on radio. In other words, the potential aggregate size of the many small markets in goods that don't individually sell well enough for traditional retail and broadcast distribution may rival that of the existing large market in goods that do cross that economic bar.

The term refers specifically to the yellow part of the sales chart at upper left, which shows a standard demand curve that could apply to any industry, from entertainment to hard goods. The vertical axis is sales; the horizontal is products. The red part of the curve is the hits, which have dominated our markets and culture for most of the last century. The yellow part is the non-hits, or niches, which is where the new growth is coming from now and in the future.

Traditional retail economics dictate that stores only stock the likely hits, because shelf space is expensive. But online retailers (from Amazon to iTunes) can stock virtually everything, and the number of available niche products outnumber the hits by several orders of magnitude. Those millions of niches are the Long Tail, which had been largely neglected until recently in favor of the Short Head of hits.

When consumers are offered infinite choice, the true shape of demand is revealed. And it turns out to be less hit-centric than we thought. People gravitate towards niches because they satisfy narrow interests better, and in one aspect of our life or another we all have some narrow interest (whether we think of it that way or not).

The Long Tail [is made possible in part because of] technologies that have made it easier for consumers to find and buy niche products, thanks to the "infinite shelf-space effect"—the new distribution mechanisms, from digital downloading to peer-to-peer markets, that break through the bottlenecks of broadcast and traditional bricks-and-mortar retail.

Box 2.3 Participatory Media Cultures

From Henry Jenkins, *Convergence Culture: Where Old Media and New Media Collide* (New York: New York University Press, 2006), 17–19.

When people take media in their own hands, the results can be wonderfully creative; they can also be bad news for all involved.

For the foreseeable future, convergence will be a kind of kludge—a jerry-rigged relationship among different media technologies—rather than a fully integrated system. Right now, the cultural shifts, the legal battles, and the economic consolidations that are fueling media convergence are preceding shifts in the technological infrastructure. How those various transitions unfold will determine the balance of power in the next media era.

The American media environment is now being shaped by two seemingly contradictory trends: on the one hand, new media technologies have lowered production and distribution costs, expanded the range of available delivery channels, and enabled consumers to archive, annotate, appropriate, and recirculate media content in powerful new ways. At the same time, there has been an alarming concentration of the ownership of mainstream commercial media, with a small handful of multinational media conglomerates dominating all sectors of the entertainment industry. No one seems capable of describing both sets of changes at the same time, let alone show how they impact each other. Some fear that media is out of control, others that it is too controlled.

Some see a world without gatekeepers, others a world where gatekeepers have unprecedented power. Again, the truth lies somewhere in between. Convergence, as we can see, is both a top-down corporate-driven process and a bottom-up consumer-driven process. Corporate convergence co-exists with grassroots convergence. Media companies are learning how to accelerate the flow of media content across delivery channels to expand revenue opportunities, broaden markets, and reinforce viewer commitments. Consumers are learning how to use these different media technologies to bring the flow of media more fully under their control and to interact with other consumers. The promises of this new media environment raise expectations of a freer flow of ideas and content. Inspired by those ideals, consumers are fighting for the right to participate more fully in their culture. Sometimes, corporate and grassroots convergence reinforce each other, creating closer, more rewarding relations between media producers and consumers. Sometimes, these two forces are at war and those struggles will redefine the face of American popular culture.

Convergence requires media companies to rethink old assumptions about what it means to consume media, assumptions that shape both programming and marketing decisions. If old consumers were assumed to be passive, the new consumers are active. If old consumers were predictable and stayed where you told them to stay, then new consumers are migratory, showing a declining loyalty to networks or media. If old consumers were isolated individuals, the new consumers are more socially connected. If the work of media consumers was once silent and invisible, the new consumers are now noisy and public. Media producers are responding to these newly empowered consumers in contradictory ways, sometimes encouraging change, some times resisting what they see as renegade behavior. And consumers, in turn, are perplexed by what they see as mixed signals about how much and what kinds of participation they can enjoy. As they undergo this transition, the media companies are not behaving in a monolithic fashion; often, different divisions of the same company are pursuing radically different strategies, reflecting their uncertainty about how to proceed. On the one hand, convergence represents an expanded opportunity for media conglomerates, since content that succeeds in one sector can spread across other platforms. On the other, convergence represents a risk since most of these media fear a fragmentation or erosion of their markets. Each time they move a viewer from television to the Internet, say, there is a risk that the consumer may not return.

Taken together these new ways of making and sharing culture have broad ramifications for the fundamental relations between production and consumption and the traditional sources of authority for culture and knowledge. By reshaping long-established standards of production and consumption, amateurs, file sharers, and bloggers challenge existing institutional and professional authority. Today we see the first glimmerings of what a fully networked public culture might look like. But persistent predictions of imminent doom for established content industries, together with fears of corporate litigation and monopolistic forces squelching the emerging common culture, indicate that the future of public culture is still very much up for grabs. Our goal in this chapter, therefore, is not to declare the forms of networked culture we describe as faits accomplis, or as inevitable forms of culture and media. Rather, we specifically selected four cases studies—music file sharing, anime fandoms, viral marketing, and news blogs—that serve as sites of contestation between the forces of government regulation, technological engineering, corporate maneuvering, and networked, viral, and laterally organized Internet groups.

Instead of setting out to write a general survey of how digital networks are changing cultural production, we are focusing on specific cases that offer alternative models with which to frame thinking about evolving relations between production and consumption. We could have chosen from many other cases, such as other fandoms, machinima, online encyclopedias, scholarly publications, as well as do-it-yourself (DIY) fashion or design movements. But we feel that the cases we present exemplify an illuminating range of dynamics in emergent networked public culture. Our first case of music file sharing is the most well-known example of the present tensions in network culture, but it represents an older form of antagonism that is currently being supplanted by new kinds of coalitions and business models based on different relationships between producers and consumers, businesses and customers, publishers and audiences. The problems that the music industry encountered in cracking down on consumer activism provided lessons for other industries, such as marketing and television, that are experimenting with new ways to reach out to fans and remixers. While new models of these relationships diffuse some of the antagonisms visible in the case of music, they also raise new questions and controversies about the role of secondary markets, the validity of knowledge, and the breakdown of common culture.

Amateur Music and Remix

The battle between the recording industry and file-sharing music fans is one crucial example of how production and consumption of cultural products is changing in the era of networked publics, and it illustrates some of the underlying issues associated with these shifts. The music industry has been revolutionized in a number of ways: networked technologies, most notably P2P applications that allow users to download and share digital music files; cheap, easy-to-use digital audio workstations and software allowing people to easily and cheaply create CD-quality music; and social software and social network platforms that have created communities of shared practice, knowledge, and expertise outside of traditional music marketing channels.

That this is a revolution for the music industry is by no means an exaggeration. From 2000 to 2006, the music industry's revenues plunged from $14.32 billion to $9.65 billion, and sales of CDs declined from 942 million to 614 million. Although sales of digital tracks in venues such as Apple's iTunes rose from virtually nothing to $1.85 billion during that time, the industry is clearly in transition (best case) or crisis (worst case).[9]

At the outset, we have to see these changes in their context. Although digital technology and the Internet gave both musicians and consumers the tools to have a massive impact on the music industry, "the biz" was far from beloved. The consolidation of record labels, the proliferation of consultant-driven radio programming and the resulting homogenization of available commercial music, the absorption and subsequent domestication of alternative music by the industry, as well as a widespread perception among the public and artists alike that the industry did not benefit musicians made both consumers and amateur musicians eager to reject it, or at least to pursue alternatives.

P2P file sharing hit the mainstream in 1999 with the release of Napster. P2P applications make files stored on a single personal computer available to other users for download over the Internet and smaller networks. Around this time the industry Big Four (Universal, Sony-BMG, EMI and Warner Brothers) accounted for approximately 80 percent of all music sales globally. The emergence of P2P was an obvious threat to corporations that perceived file sharing as theft of their intellectual property, and they acted swiftly against it. Too confidently, one industry insider declared, "we are going to strangle this baby at birth."[10] The Big Four mounted a four-fold strategy to battle P2P: extending intellectual property rights, litigating against both P2P platform providers and users, developing digital rights management restrictions, and creating a public relations campaign.[11] In a highly publicized court case, Napster was sued by the record industry and lost, leading to its demise in 2001.[12] Other file-sharing services such as Gnutella, LimeWire, Kazaa, and BitTorrent sprang up to take its place, following a more thouroughly P2P model that employed decentralized servers and thereby making them more difficult to shut down. Despite the threat of litigation, P2P is still thriving.

Although P2P sharing of music is decried by the Recording Industry Association of America (RIAA) and other industry entities that seek to protect—and expand—intellectual property laws against copyright infringement, copyleft activists argue that present intellectual laws are outdated, stifling innovation by privileging individual and corporate financial interests over the interests of the collective.[13] Media historian Siva Vaidhyanathan exemplifies this position, observing that "the copyright holder is very rarely the artist herself."[14] Others have suggested that P2P file sharing is nothing less than an act of mass civil disobedience against a corrupt industry that exploits artists.[15]

At the same time, there is the very real problem of artists being able to make a living through their music. While it can be argued that existing copyright law is not longer viable, alternative models and strategies have yet to be devised. Perhaps the market will solve the problem, as creative solutions rise and spread quickly in the new landscape. But given the legacy of intellectual property and the vast financial stakes, this will be a serious challenge.

Although the RIAA has blamed the year-over-year collapse of CD sales on P2P file sharing, some research suggests that the impact on music sales is low and that although some users avoid purchasing music due to file sharing, others may use it to sample tracks that they may later buy.[16] From this viewpoint, the music industry's troubles are somewhat self-inflicted, its decline the result of a lack of compelling new product rather than an assault by technology. Even if the jury is still out on this issue, there can be no question that the industry is in troubled times and undergoing massive changes.

Ultimately, the real threat to the established music industry is *legitimate* distribution via file-sharing applications or Web sites. These create alternatives to established distribution models, putting some level of control back into the hands of the music creators. One estimate suggests that as of 2004 the ratio of legal to illegal song downloads was 250 to 1, with billions of music files exchanged every week.[17] File sharing greatly increases the amount of available music and our ability to access it. As a result, musicians no longer need a record label to distribute their music, and fans are no longer limited to the tastes of music industry executives and retail owners. Even the industry's network of promotion and distribution has been replicated, and in some cases surpassed, by blogs, forums, and other sites that provide content producers a place to find listeners, as well as locate possible collaborators. The 2004 edition of *The Indie Bible* lists over four hundred Web sites where independent musicians can distribute their music.[18]

Apart from new means of distribution, digital technology has made it possible for musicians to produce higher quality, more sophisticated recordings, again short-circuiting the need for aid from the music industry but also changing the way that songs can be produced and giving rise to new genres. Although *sampling* (using short recorded bits of music from other sources) was used prior to the advent of digital technologies, the new tools have greatly facilitated the process to the point of being trivial from a technical standpoint. Similarly, *quoting* musical phrases has a rich history, as exemplified by jazz artists like Charlie Parker and Miles Davis, who would play variations on phrases and motifs from other material within their work. The new tools and practices move sampling and quoting to the forefront, allowing for new genres such as remix and mash-up. *Remix,* in which small portions of one song are extracted and used in a new derivative work, is an extension and expansion of sampling and quoting. *Mash-up,* which usually involves taking two or more songs and juxtaposing them, is a variation of the remix theme. Hip-hop predates digital technology, relying on old-school analog sampling techniques, but it could be argued that part of the reason for its increase in popularity and penetration in society is due to digital technology lowering the barriers for the use of samples to create new music.

Until the advent of modern digital recording technology, music had to be recorded in studios. Studio time was expensive, tape was expensive, and editing was a chore, so most musicians—apart from the wealthy and already successful—needed to have their music substantially complete before entering the studio to lay down their tracks.

In 1987 a program called Sound Tools was released as the first tapeless recording studio. Later renamed Pro Tools, this software and hardware combination created the first digital audio workstation (DAW), enabling musicians to record multiple tracks entirely on a computer and subsequently edit and play back their work. Although the sonic quality and stability of the early systems were issues, the ability to easily and *nondestructively* create, edit, and apply effects changed not only the workflow in the studio but also the creative process for the artists. By working digitally, musicians and engineers could record in a manner fitting the schedule and temperament of the artist, and editing could become a larger part of the music composition.

As computers and other digital technologies improved, studios could produce music digitally end-to-end, and independent pressing and burning of compact discs became cheaper and easier—mostly due to the availability of production houses via the Internet.[19] For musicians with access to studio time, the boundary between the professional studio and the home studio blurred: tracks could be recorded at home or in a semi-pro project studio and then taken to a larger facility for further work. Or conversely, tracks could be recorded in a professional studio (often with higher-quality microphones and better-sounding rooms) and then taken home for postproduction or further tracking of instruments not requiring precise room dynamics (e.g., electronic keyboards). The ability to swap files and project sessions with ease, either over the network or using optical media, allows musicians to move song creation and production forward without being in the same studio at the same time or even in the same time zone.

Just as P2P file sharing has undone traditional channels of distribution, the rise of the digital studio has led to the demise of many professional recording studios. In the winter of 2005 three major studios closed: Cello Studios in Los Angeles, Muscle Shoals Sound Studios in Sheffield, Alabama, and the Hit Factory in New York City.[20] Nor have things improved much lately for the studios: in 2007, Sony Studios in New York shut its doors. All were based on the traditional model of recording where record labels paid huge sums of money for studio time. Although the production technology changed, the industry did not. As *Tape Op* magazine editor Larry Crane states, "The business model of 'we have the technology needed to make records – you don't' is gone."[21] Nor is knowledge of how to run a studio exclusive anymore. Online bulletin boards and forums focused on particular software and hardware products provide users the ability to ask questions, learn new techniques, and solve problems.[22]

Beyond the economic impact on the traditional studio, home recording and mixing equipment became employed as a medium for the process and production of music as well. In particular, loop-based music production software such as ACID Pro has had a profound effect on many genres and production capabilities. Such applications allow the user to take music loops, add other sounds, and change tempos and pitch through the use of time-stretching algorithms. Since loops are digital files, users can purchase prepackaged loops on CD, download loops from Web sites, or trade loops online via P2P file sharing.

As software improved and computing horsepower increased, time/tempo algorithms became more transparent, making remix, or the bringing together of sounds from disparate recordings, not only possible but easy to accomplish. Moreover, digital music files can be easily sampled for loops and remixed to create innovative works such as DJ Dangermouse's *The Grey Album,*[23] or the Illegal Art's *Deconstructing Beck.*[24] As such works prove, music today may have no "new" content but still produce clearly original results. Moreover, the ability of someone to create music with no knowledge of how to play a musical instrument blurs the boundaries between "real" musician and amateur. It also brings issues of intellectual property to the fore, raising the question of how much sampling and remix can be done legally. Both *The Grey Album* and *Deconstructing Beck* are technically illegal, and online hosting of the music files results in cease-and-desist letters from industry lawyers.

The combination of DAW software, loop manipulation software, and the network produces many more routes for creating remixes. Some artists not only create a finished version of a song, but also make available individual instrument tracks, loops, and other sonic bits with the express goal of having people use their material in remixes.[25]

Just as music distribution and production have been reshaped by digital technology and networking, social software tools and social networking platforms have allowed individuals to act as taste-making gatekeepers for themselves and their peers. Instead of music label A&R (artist and repertoire) staff determining who gets signed and subsequently heard, social networks spread new and unknown artists to their friends and their friend's friends. As a result, MySpace has evolved as a de facto home for musicians and bands. Independent artists such as M. I. A., previously relegated to local notoriety at best, have found broad audiences through word-of-mouth and promotion on Web sites catering to niche interests. Blogs and wikis serve as venues for musicians and audiences to distribute songs and other content, while providing insight into the bands and artists. This qualitative change in the public's role in the music industry reflects emerging changes in other sectors of the culture industry. The ability of music consumers to exert increased control over what music they have access to and what they do with "their" music signals a broader shift in trends of cultural resistance, from *jamming* (where cultural products and their presumed hegemonic force are interrupted) to *poaching* (where cultural products are taken up and refashioned to convey individualized tastes and messages).[26] Or as Aram Sinnreich suggests, mash-up and remix can be thought of as movements of aesthetic resistance to the consolidation of popular music styles.[27]

Perhaps the most interesting aspect of these networked environments is the way they enable the rise of social and cultural capital.[28] Online artist, critic, and consumer often swap roles—creating, curating, and commenting for online glory rather than monetary reward. Still, the majority of online music is sold through large online aggregators such as iTunes and, until recently, almost always encoded with digital rights management (DRM) software keys to prevent sharing between users. It may well be that we are exchanging one music industry for another, although recent moves by Apple and Amazon.com to make available DRM-free music could indicate otherwise.

To be sure, music is a special case. It was the first culture industry to be threatened by the combination of low-cost digital production tools with online file sharing and social networks. Music has always been a domain of robust amateur production, making it particularly amenable to more bottom-up forms of production and distribution in the digital ecology and ripe for the disintermediation of labels and licensors. The fact that widespread music file sharing happened relatively early also meant that the existing music industry was poorly equipped to deal with the new online ecology, and it took a reactive stance rather than anticipating new practices and potential business solutions. Although the story of digital music is far from over, already it reads as a cautionary tale of the current fragility of business models built on earlier media infrastructures. P2P is a cultural economy, and as anthropologist Alfred Gell wrote, "consumption is part of a process that includes production and exchange, all three being distinct only as phases of the cyclical process of social reproduction, in which consumption is never terminal."[29]

The biggest change of all may be the reconfiguration of the status of amateur and professional. As late as 2001 the prevailing wisdom among musicians was that amateur status was "something to get *beyond.*"[30] In other words, the end game for the artist was professionalization—getting signed to a record label and following the traditional industry model. However as lines between amateur and professional blur, remix becomes embedded into culture—even beyond music—and technological changes continue to occur, "getting *beyond*" amateur status may no longer be the goal.

Transnational Anime Fandom

Fans are the lifeblood of commercial media, and yet they have often had an uneasy relationship with media industries. As enthusiasts of particular artists or series, *fandoms*—subcultures composed of fans with a common interest—can serve as the source of P2P promotional buzz as well as the base of the consumer market. But when fans cross the line into producing and trafficking in their own cultural products derived from commercial content, they create their own unique cultural forms that circulate in alternative and P2P networks

Box 2.4 Free Culture

From Lawrence Lessig, *Free Culture* (New York: Penguin Books, 2004), 7–10.

There's no single inventor of the Internet. Nor is there any good date upon which to mark its birth. Yet in a very short time, the Internet has become part of ordinary American life. According to the Pew Internet and American Life Project, 58 percent of Americans had access to the Internet in 2002, up from 49 percent two years before.[a] That number could well exceed two-thirds of the nation by the end of 2004.

My claim is that the Internet has induced an important and unrecognized change in [how culture is made]. That change will radically transform a tradition that is as old as the Republic itself. Most, if they recognized this change, would reject it. Yet most don't even see the change that the Internet has introduced.

We can glimpse a sense of this change by distinguishing between commercial and noncommercial culture, and by mapping the law's regulation of each. By "commercial culture" I mean that part of our culture that is produced and sold or produced to be sold. By "noncommercial culture" I mean all the rest. When old men sat around parks or on street corners telling stories that kids and others consumed, that was noncommercial culture. When Noah Webster published his "Reader," or Joel Barlow his poetry, that was commercial culture.

At the beginning of our history, and for just about the whole of our tradition, noncommercial culture was essentially unregulated. Of course, if your stories were lewd, or if your song disturbed the peace, then the law might intervene. But the law was never directly concerned with the creation or spread of this form of culture, and it left this culture "free." The ordinary ways in which ordinary individuals shared and transformed their culture—telling stories, reenacting scenes from plays or TV, participating in fan clubs, sharing music, making tapes—were left alone by the law.

The focus of the law was on commercial creativity. At first slightly, then quite extensively, the law protected the incentives of creators by granting them exclusive rights to their creative work, so that they could sell those exclusive rights in a commercial marketplace.[b] This is also, of course, an important part of creativity and culture, and it has become an increasingly important part in America. But in no sense was it dominant within our tradition. It was instead just one part, a controlled part, balanced with the free.

This rough divide between the free and the controlled has now been erased.[c] The Internet has set the stage for this erasure and, pushed by big media, the law has now affected it. For the first time in our tradition, the ordinary ways in which individuals create and share culture fall within the reach of the regulation of the law, which has expanded to draw within its control a vast amount of culture and creativity that it never reached before. The technology that preserved the balance of our history—between uses of our culture that were free and uses of our culture that were only upon permission—has been undone. The consequence is that we are less and less a free culture, more and more a permission culture.

This change gets justified as necessary to protect commercial creativity. And indeed, protectionism is precisely its motivation. But the protectionism that justifies the changes is not the limited and balanced sort that has defined the law in the past. This is not a protectionism to protect artists. It is instead a protectionism to protect certain forms of business. Corporations threatened by the potential of the Internet to change the way both commercial and noncommercial culture are made and shared have united to induce lawmakers to use the law to protect them.

For the Internet has unleashed an extraordinary possibility for many to participate in the process of building and cultivating a culture that reaches far beyond local boundaries. That power has changed the marketplace for making and cultivating culture generally, and that change in turn threatens established content industries. The Internet is thus to the industries that built and distributed content in the twentieth century what FM radio was to AM radio, or what the truck was to the railroad industry of the nineteenth century: the beginning of the end, or at least a substantial transformation. Digital technologies, tied to the Internet, could produce a vastly more competitive and vibrant market for building and cultivating culture; that market could include a much wider and more diverse range of creators; those creators could produce and distribute a much more vibrant range of creativity; and depending upon a few important factors, those creators could earn more on average from this system than creators do today—all so long as the RCAs of our day don't use the law to protect themselves against this competition.

Yet . . . that is precisely what is happening in our culture today. These modern-day equivalents of the early twentieth-century radio or nineteenth-century railroads are using their power to get the law to protect them against this new, more efficient, more vibrant technology for building culture. They are succeeding in their plan to remake the Internet before the Internet remakes them.

It doesn't seem this way to many. The battles over copyright and the Internet seem remote to most. To the few who follow them, they seem mainly about a much simpler brace of questions—whether "piracy" will be permitted, and whether "property" will be protected. The "war" that has been waged against the technologies of the Internet—what Motion Picture Association of America (MPAA) president Jack Valenti calls his "own terrorist war"[d]—has been framed as a battle about the rule of law and respect for property. To know which side to take in this war, most think that we need only decide whether we're for property or against it.

If those really were the choices, then I would be with Jack Valenti and the content industry. I, too, am a believer in property, and especially in the importance of what Mr. Valenti nicely calls "creative property." I believe that "piracy" is wrong, and that the law, properly tuned, should punish "piracy," whether on or off the Internet.

But those simple beliefs mask a much more fundamental question and a much more dramatic change. My fear is that unless we come to see this change, the war to rid the world of Internet "pirates" will also rid our culture of values that have been integral to our tradition from the start.

These values built a tradition that, for at least the first 180 years of our Republic, guaranteed creators the right to build freely upon their past, and protected creators and innovators from either state or private control. The First Amendment protected creators against state control. And as Professor Neil Netanel powerfully argues,[e] copyright law, properly balanced, protected creators against private control. Our tradition was thus neither Soviet nor the tradition of patrons. It instead carved out a wide berth within which creators could cultivate and extend our culture.

Yet the law's response to the Internet, when tied to changes in the technology of the Internet itself, has massively increased the effective regulation of creativity in America. To build upon or critique the culture around us one must ask, Oliver Twist-like, for permission first. Permission is, of course, often granted—but it is not often granted to the critical or the independent. We have built a kind of cultural nobility; those within the noble class live easily; those outside it don't. But it is nobility of any form that is alien to our tradition.

---

a. Amanda Lenhart, "The Ever-Shifting Internet Population: A New Look at Internet Access and the Digital Divide," Pew Internet and American Life Project, http://www.pewinternet.org/report_display.asp?r=88.

b. This is not the only purpose of copyright, though it is the overwhelmingly primary purpose of the copyright established in the federal constitution. State copyright law historically protected not just the commercial interest in publication, but also a privacy interest. By granting authors the exclusive right to first publication, state copyright law gave authors the power to control the spread of facts about them. See Samuel D.Warren and Louis D.Brandeis, "The Right to Privacy," *Harvard Law Review* 4 (1890): 193, 198–200.

c. See Jessica Litman, *Digital Copyright* (New York: Prometheus Books, 2001), ch.13.

d. Amy Harmon, "Black Hawk Download: Moving Beyond Music, Pirates Use New Tools to Turn the Net into an Illicit Video Club," *New York Times,* January 17, 2002.

e. Neil W. Netanel, "Copyright and a Democratic Civil Society," *Yale Law Journal* 106 (1996), 283.

under the radar of commodity capitalism. Fan fiction, art, music, videos, and comics are forms of long-tail media largely invisible to the mainstream but that nonetheless always existed in the shadow of commercial mass media. Now these forms of cultural production are being energized through their uptake of digital production tools and networks. Much as musical mash-ups have both celebrated and challenged the products of commercial culture, fan art, comics, and fiction have disrupted the singular authorial voice of popular novels, movies, and television.

Traditionally, commercial media make their money off the one-to-many circulation of content to mass audiences, not in the sharing of content between audiences. Activist fan groups disrupt the logic of mainstream narratives and copyright regimes, going against the grain of what Lawrence Lessig has called "permission culture—the regime of copyright restrictions that insists that all uses of copyrighted works need to be explicitly leased."[31] In the case of television, movies, and novels, the relation between fan-produced culture and commercial culture has often been a site of ongoing tension and negotiation. For example, there have been high-profile legal battles between the industry and fans of *Star Wars* and *Harry Potter.* By contrast, at least two cultural domains—anime and machinima—have been characterized by a more synergistic relationship between fan cultural production and commercial production. Here we discuss the case of anime fandoms outside of Japan, a unique but illuminating example of how fans and industry have reached some compromises in dealing with fan-produced digital media and online distribution.

Unlike music, where the means of production are relatively ready at hand, most of us do not grow up creating animated television shows as an everyday cultural practice. Even now, in an era of relatively low-cost digital animation, the level of skill and time required to produce such work is well out of the reach of amateurs. Thus, television and filmic fan production often takes the form of what Jenkins has dubbed *poaching* or what lawyers call *derivative works*—using the narrative, characters, and images from commercial media to produce other media.[32] What makes the relation between anime producers and overseas fans unique is that commercial producers, for the most part, tolerate and exploit amateur cultural production instead of ignoring it or trying to shut it down. Recognizing that activist and productive fans can create rather than detract from their business, anime producers help to circulate such collective (but commercial) imaginations.

Historically, Japanese *manga* (comics) and anime industries have taken a relatively tolerant view of fan-produced cultural content. The *doujinshi* (amateur comic) scene in Japan is enormous and has thrived since the seventies.[33] The largest convention in the country of any sort, bringing together up to five hundred thousand fans, is the biannual comic market devoted to the sale of *doujinshi.* Although these fan productions have been largely scorned by the mainstream, industry has largely tolerated it, demonstrating that, given a looser copyright regime, fan-produced derivative works can rival the mainstream commercial market in scale.

Until recently, however, in the English-speaking world, anime was a marginal form of cult media, restricted to relatively extreme fandoms that crossed over somewhat with the science fiction and fantasy world. As cultural and human traffic between Japan, the United States, and Europe increased in the eighties and nineties, the small audience for Japanese media overseas slowly began to grow.

In these early years, leaders in the fandom had some degree of communication with both the Japanese industry and its American licensors and saw their role as evangelists for anime overseas. Anime was distributed at conventions, local clubs, and via mail. Noncommercial *fansubbing* (fan subtitling) emerged, as this was the only way that English-language fans could gain access to localized versions of anime. It was during this period that anime fans began developing what Sean Leonard calls a "proselytizing commons," the free nonmarket sharing of content for the purposes of promoting and creating a new commercial market.[34] It was also during this period that fans created certain social norms about media sharing. Committed to keeping their work in the nonmarket sector, fansubbers agreed not to profit from their ventures. Seeing themselves as supporters of the anime industry, they would stop circulating their wares when a commercial English-language release was announced. Cooperation between fans and some committed overseas distributors of anime is credited with opening up the market for anime in the United States, but now that the market is relatively well-established, fansubbing is becoming more controversial.[35] Lawrence Eng describes how anime fans value intellectual property even though they engage in file sharing and the production of derivative works. They "consider it their responsibility to protect intellectual property—not just their own, but that which is created by corporate interests." He sees fansubbing as an example of "taking control of information while seeking to minimize the harm done to the producers of that information."[36]

With the advent of P2P video distribution over the Internet, the circulation of anime overseas has reached a new order of magnitude. Now, most popular anime series released in Japan will eventually be released with fansubs and distributed, via BitTorrent or on streaming video sites, to millions of fans around the world. For the most popular series, a networked group might turn around a title within a day of its release in Japan.[37] Thousands of fans watch the torrent listings or lurk on the fansub IRC (Internet Relay Chat) channels waiting for the group to give word that this week's fansub is out. As a result of this attention, anime is becoming less a niche media and more visible in mainstream media, taking over slots on popular cable channels like the Cartoon Network and becoming a mainstay of the DVD sales and rental industry. According to a recent article in *Fortune,* the output of the top U.S. DVD distributors of anime is more than the combined DVD distribution of Warner Brothers and Paramount, the two top U.S. television show distributors.[38]

In contrast to the music industry, overseas anime distribution is a case of the long tail of distribution wagging the head successfully. Rather than cracking down on fansubbers and Net distribution, the anime industry has continued to take a relatively accommodating stance, which in turn has kept organized fan groups toeing the party line. One popular fansub group, Anime-Empire, states, "We wish only to help expand the Japanese animation market to North America, without harming or impeding the business in any way. Therefore, once a title has been licensed in North America, we wish for fans to discontinue distribution of said title, and encourage others to purchase the newly released DVDs and mangas in their local anime/manga dealers."[39] There is growing diversity, however, in how both fansub groups and anime companies view fan production and distribution.

Fansubs are not the only example of fan-level nonmarket production by anime groups. Although *doujinshi* have been slower to take off outside Japan because of the effort involved, fan art, fiction, and remixed anime music videos thrive in the contemporary network ecology. Fan fiction and art have a counterpart in Japan, but anime music videos currently exist only in overseas anime fandoms that rely on digital distribution. Fans will take commercial anime footage, strip out the soundtrack, extract short clips, and edit them to conform to a song or another soundtrack (e.g., a movie trailer or advertisement) of their choosing. Often these creations are parodies of the commercial narrative or illustrate latent themes and backstories. American fans remix American media, and Japanese fans remix Japanese media. AMVs (anime music videos), however, are cultural mash-ups, localizing Japanese visual media for the different sensibilities and cultural referents of overseas fans. Anime footage set to European or American popular music is a new cultural form arising from the experiences of cross-cultural fandom. Although these are derivative works that don't depend on the craftwork of drawing and animating, even a cursory review of these productions reveals often stunning new forms of visual literacy unique to the digitally networked age. Esoteric cultural referents to anime characters and narratives are embedded in visual cues edited to conform to the audio track through lip-sync, rapid-fire cuts, and often-sophisticated, labor-intensive digital effects.

Although there are a handful of cases where anime music video creators have been asked to take their wares off the Net by corporations, these moves have rarely been initiated by the Japanese anime companies. Rather, it has been the U.S. licensors or record labels that own the soundtracks used in the mash-up videos that have been sending the cease-and-desist letters.[40]

Although the relationship between anime fandoms and commercial anime studios seems less hostile than the tension between the music industry and P2P file sharers, it is difficult to know whether we are witnessing a momentary and fragile peace or the dawn of a golden era for overseas anime fandoms during which both fan and commercial distribution will continue to flourish. As the market for anime overseas becomes increasingly established, anime industries may feel that they don't need fans to evangelize their works, and break from their historical tolerance of fan production and distribution. Larger audiences and fandoms also mean a less tight-knit community that might lack the discipline to police themselves. The case of the transnational circulation and remix of anime provides hints as to some possible futures for networked publics in which amateur remix and derivative works will be tolerated; whether this model survives what seems to be an inevitable scaling up and scrutiny by mainstream media remains to be seen.

Viral Marketing

More willingly than music or anime, marketing embraces networked publics to harness the power and influence of the group once known as consumers. Now that emerging technologies have splintered audiences into micro-niches, the era of demographics-driven campaigns is widely considered to be over. In this fragmented media landscape, marketers are ever more dependent on fans to spread the word. Viral marketing assumes consumers, not firms, have the most influence in the creation of brands.[41] Increasingly marketers attempt to tap into fan culture to co-opt fans' creativity for relatively inexpensive grassroots marketing campaigns. From the point of view of marketers, fans can serve as brand evangelists, essential partners in negotiating a product's meaning in the constant conversation that is native to networked publics.

According to Henry Jenkins, ever since Napster popularized file sharing, the approach to new-media fandom has split along two general lines. The film, television, and recording industries have predominantly attempted to regulate fan engagement with their products, while Internet and games companies have been more willing to experiment, adopting an approach that enlists fans in the work of content production and brand promotion. Jenkins refers to these two models as *prohibitionist* and *collaborationist*.[42] According to Jenkins, the former will fail to accommodate network demand for participation, one of the key products of the new media market, and thus lose fans to more tolerant forms of media. If the relationship with fans is becoming increasingly significant in the networked era, the role of marketing in mediating between producers and consumers will change. The way that marketers adapt this collaborationist approach to creating campaigns in a deeply fragmented media landscape suggests possible future strategies by which other media industries will engage networked publics.

A variety of disruptive technologies allow consumers to customize their media by choosing more selectively from a wider array of sources and time-shifting their consumption patterns. Traditional marketing practices are threatened by technologies such as set-top boxes, video on demand, and podcasting that allow consumers to cut ads from media. This results in a transformation in the media landscape, moving it from a push to a pull ecology, from a condition in which consumers passively receive content to one in which consumers begin to set the terms of their engagement. Rather than spending their entire marketing budget on thirty-second spots that dwindling audiences passively receive, marketers are increasingly interested in producing experience-driven campaigns, a phenomenon of convergence in which New York advertising meets Hollywood entertainment, what *Advertising Age* editor Scott Donaton refers to as "Madison and Vine."[43]

Social networking technologies, from e-mail to MySpace, have given consumers the power to transform brands. Eager to channel this participation, while still wary of grassroots criticism that could spiral out of control, marketers are attempting to create fan-driven experiences adapted to a wide variety of media. Viral marketing assumes consumers, not firms, have the most influence in creating brands.[44] Using social networks to spread the word, viral media grew as an epiphenomenon of e-mail forwarding, which according to Dan Brooks (famous for his spoof Volkswagen-suicide bomber advertisement), echoes a tacit understanding in the age-old practice of telling jokes: "If you repeat it, you own it."[45] But activist brand detractors can also get into the game, appropriating brands to transmit their own messages. Take the instance of a series of Nike sweatshop e-mails initiated by Jonah Peretti. After Nike responded over e-mail that he could not customize his shoes with the word *sweatshop,* Peretti forwarded the e-mail correspondence to friends. The e-mail subsequently spread virally, becoming an Internet phenomenon and eventually landing Peretti a spot on the *Today Show.*[46]

Media executive Jim Banister provides a useful theoretical frame for viral social networking. To describe such ventures as eBay and Friendster, Banister uses the term *enginet* to refer to an algorithmic structure that combines code, form, and function to create community-driven experiences in which the users themselves have found innovative, often unanticipated, ways to connect with one another. According to Banister, the successful enginet pulls visitors seamlessly through a variety of states, from producer to distributor to marketer to vendor to consumer. While Banister locates the antecedent of the enginet in the value-chain marketing schemes of Avon and Mary Kay, he claims that the frictionless nature of networked media has exponentially scaled them into entire ecosystems.[47]

Enginets use shared-judgment systems to create reputation-based "value nets" that Banister says leverage a complex combination of community impulse, egocentrism, and individual superego with its desire to judge. The tension of these traits has produced the bizarre category of Internet fame, often shamelessly lowbrow, such as in the example of a popular thirteen-year-old video blogger named Bowiegirl, whose fame appears to be as much the result of the mockery she receives as of any admiration. Even so, Bowiegirl became an unintentional spokesperson for Logitech after having featured one of their webcams in a late-night bedroom confessional.[48]

Although brand enthusiasts and detractors seem to be growing more and more empowered, marketers are ill at ease letting their reputations be determined by amateurs. It has become commonplace, however, for marketers to work from within the viral space, by creating campaigns cleverly dressed down in the aesthetics of amateur cultural production. The FX channel, for example, created a MySpace profile for a fictional character from their television program *Nip and Tuck* in order to promote the show. The pioneers of the fictive technique have been video game marketers and they continue to push forward, using fake blogs to seed elaborate online hoaxes. Working with the marketing firm Wieden + Kennedy, the game developer Sega created a viral campaign for the release of their game ESPN NFL Football 2K4 that passed itself off as a legitimate amateur homepage by a game tester named Beta-7. The imaginary tester claimed the game made him black out and fly into uncontrollable fits of rage. The phony site featured supposedly leaked confidential memos of a cover-up by Sega, which claimed they had knowledge of the health hazards of the game, as a way of appealing to extreme gamers.[49] In the world of the enginet, it seems that marketers are increasingly coming to resemble political spin doctors, carefully leaking disinformation to the press in order to advance an agenda, thwart detractors, and manipulate public opinion.

Media theorist Holly Willis proposes two categories for viral media: those that are "simply unseemly and outrageous," such as Brooks's Volkswagen ad, and those that "leave you very unsure about what you're viewing."[50] The majority of successful viral video clips conform to the former category, the most successful being Crispin Porter & Bogusky's Subservient Chicken Web site, a satire of online webcam pornography, developed for Burger King's "Have it Your Way" campaign. Featuring a database of video clips of a man in a chicken costume, the Subservient Chicken would respond to commands from site visitors and was ultimately responsible for driving one in six visitors to Burger King's main site.[51] Falling more clearly into the latter category is the emerging genre of alternative reality gaming (ARG). ARGs create entire self-contained worlds on the Web, often comprising a vast array of assets—logos, photos, scripts, movies, audio recordings, corporate blurbs, graphic treatments, flash movies—embedded within a network of (untraceable) Web sites. Involving a variety of complex puzzles, marketing experiences such as "the Beast," developed to promote the Spielberg film *AI,* take several weeks or months to solve and are far too complex to be solved by a single player. Networked audiences work together to process a huge amount of story information, building a collaborative relationship with each other as well as with the brand. Used with great success to market the film the *Blair Witch Project* (a film with a budget of $35,000 that grossed over $248 million), this technique is increasingly being used to market video games.

When describing the medium of ARGs, fans often note that the best-designed experiences explicitly blur the lines of reality.[52] Though an undeniably powerful new medium uniquely adapted to the multimedia context of the Web, ARGs as hype-machines could also prove somewhat treacherous territory for marketers, as the online consumer is increasingly sensitive to being manipulated and increasingly adept at exposing deceptive practices. For example, Cillit Bang, a UK cleaning product brand, was forced to publicly apologize for conducting a deceptive viral marketing campaign in which members of its marketing team posed as fictional characters on the Web to place thinly disguised ads. The campaign unraveled when the marketers were exposed by bloggers.[53]

As the forces of media disruption proliferate and audiences are increasingly lured away from official distribution channels, marketers must either adapt to the networked environment and redefine their relationship with consumers or become irrelevant. When describing the medium, ARG fans will often invoke the ideal of TINAG (This Is Not A Game), as the best of these experiences are explicitly intended to blur the lines of reality.[54]

Such developments will not be lost on marketers. They will have to adopt a view of the entire field of cultural production in order to successfully invite people to participate in constructing compelling marketing "experiences." As the relationship evolves between production and consumption, Jenkins maintains there must be detente between political economy and audience research.[55]

Perhaps we will find that, as in nature, mutualism and parasitism are, in fact, not discrete categories but rather a continuum of interaction. By creating a public arena shared by both nonmarket amateurs and commercial professionals, the Internet makes the engagements between these different parties necessarily more intimate.

Online News

In the case of online news, the relationship between the commercial industry and DIY producers is less contentious than it is in some of the other cases surveyed here. With increasing opportunities for amateur cultural production, it is clear people are actively resisting the content and practices of mainstream news, partly by using it as a launching pad to offer contesting points of view and alternative practices. Evolving digital communication tools and practices are clashing with those of traditional news media, resulting in paradox and contradiction. Stories filed by so-called embedded reporters in Iraq, for example, are being trumped by personal e-mails and photos from soldiers; Western-trained journalists in middle eastern countries are being criticized for lacking professionalism while Western audiences surf to Arab outlets to get news absent from Western reports; bloggers are working out tacit ethical codes for themselves while editorial opinions leak into all aspects of mainstream news publishing and programming.

Echoing these contradictions is the fact that one of the central assumptions about the news—its tie to democracy—grows more complex each day. On the one hand, scholars such as Robert McChesney, Edward Herman, and Cass Sunstein see civic culture as deteriorating, the flow of information and opinions limited by media consolidation, various forms of self and government censorship, and the fragmentation of audiences.[56] On the other hand, scholars such as Benkler and Jenkins celebrate DIY media for expanding the ranks of informed citizenry and facilitating the development of an engaged and participatory transnational culture.[57] Benkler suggests that the network, with its "variation and diversity of knowledge, time, availability, insight, and experience as well as vast communications and information resources," has taken over the watchdog function of the press, making it irretrievably a peer-to-peer activity.[58] Many analysts, faced with the complexities of the networked news environment, have simply divided the landscape into two spheres, new and old media, pitting them against each other. But it is increasingly evident that the landscape grows more fully integrated every day. If news industry professionals are acting on this point reluctantly, media consumers-turned-producers have recognized it instinctively for some time.

The balance of power between news providers and news consumers has shifted. Web publishing tools and powerful mobile devices, combined with an increasing skepticism toward mainstream media, have prompted readers to become active participants in the creation and dissemination of news. Video and text bloggers, DIY media activists, and professional journalists are struggling over the right to define the truth and to determine what form and practice of news production yields more credible product. Is credibility the domain of elite media institutions that abide by professional codes? Or do bloggers, with their editorial independence, collaborative structure, and merit-based popularity more effectively inform the public? The truth, as the exclusive domain of authorities and the journalists who use them as sources, is receding, making way for communication created by the public and based on peer-produced and -distributed information, storytelling, and exchange. With this shift come anxieties. The news industry focuses on the viability of its business model and the sustainability of its products. Some analysts of civic culture question how the public will get the information it needs to participate as citizens, concerned that the individualized new-media environment will serve less to weave society together than to break it apart.[59]

An era with millions of specifically tailored informational pods is viewed by analysts both as a liberation of democracy and as a horror of narcissistic isolation. New-media networks may well provide a platform where all voices can be heard, but not all voices attract equal amounts of attention. A small set of so-called A-list bloggers garner the majority of blogosphere traffic. In his essay "Power Laws, Weblogs, and Inequality," Clay Shirky argues that these inequalities are not a failure of the system but rather an inevitable side effect of freedom of choice: "In any systems where many people are free to choose between many options, a small subset of the whole will get a disproportionate amount of traffic (or attention, or income), even if no members of the system actively work toward such an outcome."[60] As the chapter on politics observes, for some, this star system is evidence that digital networks reflect offline power dynamics, while for others the merit-based process by which bloggers achieve star status is still an improvement over the previous status quo in which big media dominates.

Despite millions of dollars spent on high-profile online editions, mainstream news outlets have been reluctant to fully embrace the possibilities of digital technologies. Most traditional news organizations offer only the illusion of online interactivity, participation, and collaboration. In the spring of 2006, for example, the *New York Times* debuted the first remodeling of its Web site in more than five years. The new site emphasized personalization and something site editor Len Apcar called "lean-in" design, which aimed to get readers "to read and click and keep clicking and dig deeper into the site."[61] Coaxing readers to "lean-in" and to click more deeply into the *New York Times* news product, however, is very different than getting the reader involved in the news production. Encouraging comments and analysis, fostering contributions of reporting and fact-checking, or asking readers to weigh in on and help shape the news agenda is what truly interactive news Web sites, such as those run by Current TV, Digg, and NewsTrust, are designed to do. Nor is the distance maintained between media outlets like the *Times* and their audiences necessarily a guarantor of authority. In the same years interactive news sites were developing across the Web, for example, Judith Miller's controversial and inaccurate reporting on the existence of weapons of mass destruction in Iraq spurred strong responses from readers that mostly never reached Miller or her editors. Messages that got through failed to influence editorial policy toward Miller and her writing.[62] She remained a loose cannon in the newsroom, and when her stories were exposed, the *Times* brand was tarnished by accusations of insiderism and isolation.[63]

Traditional news media are using the Internet as a new distribution channel, experimenting in so-called conversational digital features, but they are not reconfiguring their fundamental stance toward journalistic authority and authoring conventions. Reporter blogs are now a staple among newspaper and broadcast networks looking to heighten the engagement of their readers. The *Houston Chronicle* online and British newspaper the *Guardian's* "Comment is Free" sections, for example, have been lauded for the quality of their design, content, and innovation.[64] According to Lisa Stone of BlogHer.com, "Newspaper blogs that work are carefully planned, openly executed exercises in public conversation about news and information. These blogs allow comments and turn into 24/7 town hall meetings about everything from the headlines to how well the paper is doing to deliver and discuss the news." In her view, successful newspaper blogs are an extension of their op-ed pages.[65] Those that don't blog well, according to Bob Cauthorn of CityTools.net are "simply spilling more of the same voices onto the public streets."[66] He argues that even the best staff-written blogs do not diversify the news content because they rarely elicit reader comments, and when those few comments turn hostile even the most committed organizations turn them off.[67] Satirical newsman Jon Stewart echoed this sentiment when he described mainstream media blogs as "giving voice to the already-voiced."[68]

Resistance to the full participatory potential of new media is defended by the industry on the grounds that, first, effective investigation, particularly on an international scale, requires resources and a certain amount of organizational and political clout. Second, industry defenders maintain that only corporate media have the financial resources to stand up to government and other corporate organizations in upholding the public interest.[69] Advocates of emerging forms of journalism, however, argue that collaboration is a resource more valuable than institutional backing in both cases.[70] Benkler refers to the network reportage that exposed the inadequacy and corruptibility of Diebold Election Systems voting machines.[71] The P2P Diebold investigation is a compelling example of the potential of networked journalism. After bloggers condemned it for inaccuracies, the voting system was partly decertified in California, and voting machine policy was altered as a result in several states.

The terms of debate, however, lag behind the experience of news information as it is created and received. During the 2005 riots in France, for example, people involved in the story used new media in sophisticated ways and the lines between participants, reporters, and audiences grew porous. The Web was by far the most dynamic source of information of every kind, a flood of images, stories, podcasts, video, critiques, corrections, and metanarratives. During the riots, mainstream outlets rushing to keep up mimicked participatory formats on their sites. Reporters and editors grazed the Web as a way of generating content and adapting new technologies. The French daily *Liberation* and the Swiss weekly *L'Hebdo,* among other professional media outlets, used blogs as an essential aspect of their coverage. *Liberation* promoted its blog as an up-to-the-minute, wire-style stream of information, whereas *L'Hebdo* used its blog to post in-depth analyses by reporters sent to act as participant-observers on a rotating basis to Bondy, one of the northern Paris suburbs close to where the rioting began. The public was encouraged to comment and excerpts of the reaction were published in its printed news weekly. *L'Hebdo* editors later passed the keys of the blog on to the inhabitants of Bondy by sending aspiring youth from the suburb to Lausanne for a week-long training program.

In announcing the program, *L'Hebdo* acknowledged the irony of what it referred to as the "Bondy Blog Academy," a thinly veiled effort to diversify the news while exploiting Bondy youth for bloggy content—that is, content prized for seeming to be diverse and unfiltered. The Bondy bloggers had no journalism training other than the extemporized *L'Hebdo* academy but gained immediate access to major news audiences.[72]

Network discourse about the riots was equally influenced by the mainstream agenda. Bloggers responded to questions raised in the newspapers and on TV and commented on mainstream coverage or politician responses to the unrest. Banlieue-dweller and gamer Alex Chan made a machinima film on the riots he titled *The French Democracy,* featuring pre-rendered New York City sets and characters and English-language subtitles, an example of the kind of cultural mash-up that characterizes the current transnational media ecology.

Distinctions between new and established media were also used to convey the story. Activists hacked the official Web site of Clichy-sous-Bois, where the riots originated, and posted a fake press release reporting the resignation of the mayor—a protest technique increasingly used by the antiglobalization movement and other activist groups, which likewise create ersatz news broadcasts, mock press releases, and phony corporate Web sites.[73] While listservs and blogs have long used mainstream media as a springboard for critique and discussion, now we see media activists mixing political critique with the tools and idioms of entertainment media, mobilizing hybrid cultural genres that challenge dominant cultural norms and mainstream media coverage and agenda setting.

Text-based Web sites and blogs have proliferated rapidly in part because text is easily produced. It is also the case that the widespread emergence of DIY journalism, which often depends on direct poaching of mainstream news products, has not been met with the same contestation over intellectual property that is occurring in music and other creative industry sectors. Rather than trying to shut down online news, mainstream outlets are making attempts to adopt a more collaborative and viral model in part by poaching DIY products, practices, and at times, values. In embracing key characteristics of network communication, however, especially interactivity, journalists will have to partly surrender authority, what media scholar Mark Deuze calls the "we write, you read" dogma of modern journalism.[74]

To many in the industry, positioning traditional news practices and products in the new-media environment is a rabbit hole. If mainstream outlets are neither first on the scene with breaking news, nor have the authority to deem what is news and what is truth, what do they have to sell? Much as music fans and game hackers are reconfiguring corporate entertainment media, DIY online news, by depending upon critiques the "news from the core," is supplementing and altering contemporary news as product, information, and experience.

Conclusion

The future of networked public culture is contested. The only thing we can be sure of is that the forms it will take will be highly variable. If there is a general trend toward more outspoken, unruly, and mobilized publics, the specificities of how networked culture plays out in particular arenas is highly dependent on media type, industry make up, infrastructures, geopolitics, and cultures of consumption and production. Battle lines were drawn early in the showdown between P2P and commercial distributors of music and are just now beginning to soften along some boundaries surrounding amateur music. An industry that tried to maintain the status quo wound up facing double-digit declines in revenue year after year. In contrast, anime provides a countercase of historical synergies between consumers and producers, in an industry that is just now starting to flex its muscles in the global arena. Similarly, marketing and advertising, so responsive to each shifting tide in public behavior and whim, sniff out trends and mimic styles from the counterculture even as they seek to reign in and channel these viral energies in ways that consolidate the corporate bottom line. The flexibility of marketing media contrasts with the limits of professional news media. Where the former prizes innovations in form and content almost by definition, the latter seeks legitimacy through standardization and consistency. Although news organizations are attempting to update their products and practices, they are tied to structures of authority and professionalism and to commitments as arbiters of the public sphere. Despite the turn toward news as entertainment, there are resilient, principled investments in maintaining the separation between journalism and opinion, newspapers and blogs that go to the heart of the norms we continue to use to assess authority, fact, and credibility.

Networked cultural production assails traditional structures of authority and disrupts the received logic of consumption by breaking down barriers between consumers and producers. In the cultural genres outlined previously, the public, formerly seen as an audience, is now integral to the process of production and distribution, regardless of the extent to which their power to shape the process has been accepted and integrated by existing authorities. Although networked music fans met fierce resistance from the recording industry, they have profoundly influenced music itself, reordering production and distribution in ways that have expanded understandings across genres. Definitions of the most basic terms—song, songwriter, musician, performance—have changed. Likewise, anime fans, enjoying a mostly synergistic relationship with commercial producers, add layers of meaning and popularity to industry products by remixing, adapting, and localizing them. And where the advertising industry is embracing what some of its leaders view as the connective chaos of the network by using individual consumers or agency-created consumer avatars to push products into the depths of digital social networks, the news industry seems to be entrenching itself into a smaller but still bounded domain. Even as the definitions of journalism fade on all sides and the news environment expands over cultural, national, and genre borders to every corner of the blogosphere and beyond, the industry is reluctant to let its audience in.

Although our voice throughout this chapter is mostly a celebratory one, cheering on the emergent energies of a public that had been mostly invisible in the age of mass media, dangers remain. We are still at the beginning of the trajectory toward lateral networking of public culture. The change will be more additive or accommodating than a coup d'etat. Each industry, each medium, and each fandom will need to find its own point of longer-term stability, which is likely to include a somewhat chastened, though still powerful, commercial media apparatus. The standards of authorial voice, professional artistic vision, and journalistic integrity are cultural values that we are not likely to abandon entirely, even though we may welcome a louder voice of critique and remix from diverse publics.

More significant, however, than the compromises of the culture industries are the shifts in cultural referents and creative form that are on the horizon. Convergence culture is not only a matter of industry and technology but also more importantly a matter of norms, common culture, and the artistry of everyday life. Professional commercial media brought us a slick common culture that has become a fact of life, the language of current events, shared cultural reference, and visual recognitions that lubricate our everyday interactions with one another. Commercial media provide much of the source material for our modern language of communication. The current moment is perhaps less about overthrow of this established modality of common culture, and more about adding a new set of communicative and expressive modes to the mix. At best, this is about folk, amateur, niche, and nonmarket communities of cultural production mobilizing, critiquing, and remixing commercial media to creatively produce new cultural forms. At worst, this is about the fragmenting of common culture or the decay of shared standards of quality, professionalism, and accountability.

The history of networked public culture has opened with a narrative of convergence and participatory culture; we lie at the crossroads of multiple unfolding trajectories.

Notes

1. Arjun Appadurai and Carol Breckenridge, "Why Public Culture," *Public Culture* 1, no. 1 (1998): 5–9.

2. Henry Jenkins, *Textual Poachers: Television Fans and Participatory Culture* (New York: Routledge, 1992).

3. Benkler, *The Wealth of Networks: How Social Production Transforms Markets and Freedom* (New Haven: Yale University Press, 2006).

4. Hagel and Brown, *The Only Sustainable Edge: Why Business Strategy Depends on Productive Friction and Dynamic Specialization* (Cambridge: Harvard Business School Press, 2005).

5. Lovink "Blogging, The Nihilist Impulse," *Eurozine,* http://www.eurozine.com/articles/2007-01-02-lovink-en.html.

6. Benkler, *The Wealth of Networks,* 93–95.

7. Jenkins, *Convergence Culture* (New York: New York University Press, 2006), 44–92.

8. Anderson, "The Long Tail," *Wired* 12, no. 10 (October 2004): 170–177, http://www.wired.com/wired/archive/12.10/tail.html.

9. Joseph Palenchar, "Digital Surge Fails to Stop Music Industry Decline in '06," *Twice: This Week in Consumer Electronics,* April 19, 2007, http://www.twice.com/article/CA6435083.html?nid=2402.

10. Martin Kretschmer, George Michael Klimis, and Roger Wallis, "Music in Electronic Markets," *New Media and Society* 3, no. 4 (2001): 426.

11. Andrew Whelan, "Do U Produce? Subcultural Capital and Amateur Musicianship in Peer-to-Peer Networks," in *Cybersounds: Essays on Virtual Music Culture,* ed. Michael D. Ayers, 57–81 (New York: Peter Lang Publishing, 2006).

12. The current Napster has nothing to do with this original incarnation, as Roxio bought the rights to the name following Napster's bankruptcy.

13. See, for example, Creative Commons at http://www.creativecommons.org, and Lawrence Lessig, *Free Culture: How Big Media Uses Technology and the Law to Lock Down Culture and Control Creativity* (New York: The Penguin Press, 2004).

14. Carrie McLaren, "Copyrights and Copywrongs: Interview with Siva Vaidhyanathan," *Stay Free!* no. 20, http://www.stayfreemagazine.org.

15. Downhill Battle, "Civil Disobedience, p2p," http://www.downhillbattle.org/?p=341.

16. Derek Caney for Reuters, "CD Sales Down," DMusic.com, August 26, 2002, http://news.dmusic.com/article/5340; Felix Oberholzer and Koleman Strumpf, "The Effect of File Sharing on Record Sales: An Empirical Analysis," http://www.unc.edu/~cigar/papers/FileSharing_March2004.pdf.

17. Fraser MacDonald, "Downloader's Guide to MP3," *Stuff,* June 2004, 48–59.

18. David Wimble, *The Indie Bible* (Ottawa: Big Meteor, 2004).

19. Louise Meintjes, "Review of *Representing African Music: Postcolonial Notes, Queries, Positions,* by Kofi Agawu" *Journal of the American Musicological Society* 59, no. 3 (2006): 769–777.

20. Jonathan Sterne, "On the Future of Music," *Cybersounds,* 255–263.

21. Crane, "It's the End of the World as We Know It (and I Feel Fine)," *Tape Op,* March/April 2005, 82–91.

22. For example, see Digidesign User Conference, http://duc.digidesign.com; BigBlueLounge.com, http://www.bigbluelounge.com/forums/index.php. The RecPit

forums at http://www.prosoundweb.com/recpit/ were particularly noteworthy, often hostile, generally entertaining, but are now defunct.

23. Michael D. Ayers, "The Cyberactivism of a Dangermouse," *Cybersounds,* 127–136.

24. Illegal Art, http://www.illegal-art.org. For a critical perspective, see Steven Shaviro, "Illegal Art: Deconstructing Beck," *ArtByte,* June–July 1998, 16–17.

25. See Simple Flower, http://simpleflower.com/remix.php.

26. Henry Jenkins, *Convergence Culture.*

27. Sinnreich, "Mash it up! Hearing a new musical form as an aesthetic resistance movement," http://aramsinnreich.typepad.com/aram_squalls/remix_culture/index.html.

28. Andrew Whelan, "Do U Produce?" *Cybersounds,* 255–263.

29. Gell, "Newcomers to the World of Goods: Consumption among the Muria Gonds," in *The Social Life of Things,* ed. Arjun Appadurai, 110–38 (Cambridge: Cambridge University Press, 1986).

30. Robert Drew, *Karaoke Nights: An Ethnographic Rhapsody* (New York: AltaMira, 2001).

31. Lessig, *Free Culture: How Big Media Uses Technology and the Law to Lock Down Culture and Control Creativity* (New York: The Penguin Press, 2004).

32. Jenkins, *Textual Poachers: Television Fans and Participatory Culture.*

33. Sharon Kinsella, "Japanese Subculture in the 1980s: Otaku and the Amateur Manga Movement," *Journal of Japanese Studies* 24, no. 2 (1998): 289–316.

34. Leonard, "Celebrating Two Decades of Unlawful Progress: Fan Distribution, Proselytization Commons, and the Explosive Growth of Japanese Animation," *UCLA Entertainment Law Review* 12, no. 2 (2005).

35. Jordan Hatcher, "Of Otakus and Fansubs: A Critical Look at Anime Online in Light of Current Issues in Copyright Law," *Script-ED* 2, no. 4 (2005): 514–542.

36. Eng, "Otaku Engagements: Subcultural Appropriation of Science and Technology," (Ph.D. dissertation, Rensselaer Polytechnic University, 2006), 97–99.

37. See Hatcher, "Of Otaku and Fansubs," for an overview of the digital fansubbing process.

38. Daniel Roth, "It's . . . Profitmón!" *Fortune,* December 12, 2005, 100.

39. Anime-Empire, "Mission Statement and History," http://www.anime-empire.net/anime-empire/about.php.

40. This overview of AMVs is based on Mizuko Ito's ongoing research on anime fandom.

41. Paul Berry, "Paul Berry's Viral Marketing Advice," G4 *Attack of the Show!* blog, http://www.g4tv.com/attackoftheshow/blog/post/430607/Paul_Berrys_Viral_Marketing_Advice.html.

42. Jenkins, *Convergence Culture,* 201.

43. Donaton, *Madison & Vine* (New York: McGraw-Hill, 2004).

44. Berry, "Paul Berry's Viral Marketing Advice," http://www.g4tv.com/attackoftheshow/blog/post/430607/Paul_Berrys_Viral_Marketing_Advice.html.

45. Holly Willis, "Ad Libs: The virulent strain of commercials infecting the Web," *LA Weekly,* February 17, 2005, http://www.laweekly.com/film+tv/screen/ad-libs/934.

46. "The Life of an Internet Meme: A History of the Nike E-mails," shey.net, http://www.shey.net/niked.html.

47. Banister, *Word of Mouse: The New Age of Networked Media* (Los Angeles: Agate Publishing, 2004).

48. Greg Sandoval, "YouTube's 'Bowiechick' and the Spiders From Marketing," CNET.com.au, http://www.cnet.com.au/software/internet/0,39029524,40061704,00.htm.

49. Max Lenderman, *Experience the Message: How Experiential Marketing is Changing the Brand World* (New York: Carroll & Graf, 2005).

50. Willis, "Ad Libs," http://www.laweekly.com/film+tv/screen/ad-libs/934.

51. Mae Anderson, "Dissecting 'Subservient Chicken,'" *Adweek,* March 07, 2005, http://www.adweek.com/aw/national/article_display.jsp?vnu_content_id=1000828049.

52. Dave Szulborski, *This Is Not a Game: A Guide to Alternate Reality Gaming* (Rochester, NY: Lulu.com, 2005).

53. Bobbie Johnson, "Cleaner Caught Playing Dirty on the Net," *The Guardian,* October 6, 2005, http://technology.guardian.co.uk/online/story/0,16545,1585731,00.html.

54. Szulborski, *This Is Not a Game.*

55. Henry Jenkins, "The Cultural Logic of Media Convergence," *International Journal of Cultural Studies* 7, no. 1 (2004): 36.

56. McChesney, *The Problem of the Media: U.S. Communication Politics in the Twenty-First Century* (New York: Monthly Review Press, 2004); Herman and Chomsky, *Manufacturing Consent: The Political Economy of the Mass Media* (New York: Pantheon, 2002); Sunstein, *Republic.com* (Princeton: Princeton University Press, 2002).

57. Benkler, *The Wealth of Networks;* Jenkins, *Convergence Culture.*

58. Benkler, *The Wealth of Networks,* 264.

59. Sunstein, *Republic.com.*

60. Shirky, "Power Laws, Weblogs and Inequality," Clay Shirky's Writings About the Internet, http://www.shirky.com/writings/powerlaw_weblog.html.

61. Robert Niles, "The Gray Lady Weaves a New Website," *USC Annenberg Online Journalism Review,* April 9, 2006, http://www.ojr.org/ojr/stories/060409niles/

62. Byron Calame, "The Miller Mess: Lingering Issues Among the Answers," the *New York Times,* October 23, 2005, http://www.nytimes.com/2005/10/23/opinion/

23publiceditor.html?ex=1183521600&en=97a44ee32a98a215&ei=5070; Also see Michael Massing and Orville Schell, *Now They Tell Us: The American Press and Iraq* (New York: DEL-New York Review Books, 2004), originally published in the *New York Review of Books* 51, no. 3 (February 26, 2004); Robert G. Kaiser, Judith Miller, James Risen, Reply by Michael Massing "'Now They Tell Us': An Exchange," *New York Review of Books* 51, no. 5 (March 25, 2004).

63. Massing and Schell, *'Now They Tell Us.'*

64. Jay Rosen, "The Best Blogging Newspapers in the U.S.," PressThink blog, http://journalism.nyu.edu/pubzone/blueplate/issue1/best_nwsps/; Jeff Jarvis, "Criticism is Free," Buzz Machine blog, http://www.buzzmachine.com/2006/11/02/criticism-is-free/.

65. Robert Niles, "Can Newspapers do Blogs Right?" *Online Journalism Review,* (April 23, 2006), http://www.ojr.org/ojr/stories/060423niles/.

66. Niles, "Can Newspapers do Blogs Right?"

67. Niles, "Can Newspapers do Blogs Right?"

68. Media Guerrilla, "Jon Stewart/Daily Show on Watching The Watchers," Media Gorilla blog, http://www.mguerrilla.com/media_guerrilla/2005/05/jon_stewartdail.html.

69. Neil Netanel, "The Commercial Mass Media's Continuing Fourth Estate Role," in *The Commodification of Information,* ed. Neil Weinstock, Nick Netanel, and Niva Elkin-Koren (London: Kluwer Law International, 2002).

70. Jay Rosen, "Introducing NewAssignment.Net," PressThink blog,http://journalism.nyu.edu/pubzone/weblogs/pressthink/2006/07/25/nadn_qa.html.

71. Benkler, *The Wealth of Networks,* 237–245.

72. Alain Jeannet, "Bondy Blog Academy," Le Bondy Blog, http://previon.typepad.com/hebdo/2006/02/bondy_blog_acad.html.

73. See, for example, RTMark, http://www.rtmark.com.

74. Deuze, "Online Journalism: Modeling the First Generation of News Media on the World Wide Web," *First Monday* 6, no. 10, http://www.firstmonday.org/issues/issue6_10/deuze/#d2.

3

Politics: Deliberation, Mobilization, and Networked Practices of Agitation

Merlyna Lim and Mark E. Kann

During the decade and a half since the release of Mosaic, the first viable Web browser, the Internet has gone from being a promising platform for politics to an integral part of daily political life. Even before the Web, during the 1992 presidential elections, third-party presidential candidate and technology entrepreneur Ross Perot called for "electronic town halls" allowing citizens to communicate directly with elected officials as a corrective to the well-funded special-interest groups that he saw pervading politics.[1] In fall 2007, the two-year-old video-sharing site YouTube teamed with CNN (founded in 1980 as the "Cable News Network" but also one of the most popular online news sites) to cosponsor Democratic and Republican presidential debates in which citizens submitted questions to the candidates through videos they produced themselves and uploaded to the site. For David Bohrman, chief of CNN's Washington bureau, this new model was "the most democratic of all possible structures," facilitating direct dialogue between politicians and the public.[2] During the Democratic debate, a snowman asked candidates about his future in a world of global warming, a lesbian couple challenged candidates to answer whether or not they would one day be able to get married, and a man holding a gun that he called his baby inquired into candidates' positions on gun control. Although the responses were business-as-usual, the YouTube debates emphasized that the Internet had become part of mainstream politics.

Today, both in the United States and other countries, it is common for citizens, candidates, political parties, fund-raisers, consultants, lobbyists, interest groups, legislators, and bureaucrats to have online strategies for advancing their goals. E-government literature is saturated with suggestions on how individuals, groups, and officials can communicate and compete more effectively, improve job performance, enhance public services, strengthen legitimacy, and heighten impact. From producing videos for dissemination primarily on You-Tube to maintaining active campaign blogs, politicians and government officials on the Left and Right alike use the Internet to spread their message.

But has the Internet transformed politics in any way? Perot was by no means alone when he suggested the Internet would breed a more democratic political culture. In 1993, Mitch Kapor, cofounder of renowned online civil rights nonprofit, the Electronic Frontier Foundation, called for a new "Jeffersonian Ideal . . . a system that promotes grassroots democracy, diversity of users and manufacturers, true communications among the people, and all the dazzling goodies of home shopping, movies on demand, teleconferencing, and cheap, instant databases . . . composed of high bandwidth, an open architecture, and distributed two-way switching."[3] A year later, Al Gore described his vision of the Internet as a "Global Information Infrastructure" that would "allow us to share information, to connect, and to communicate as a global community." In essence a "distributed, parallel computer . . . a metaphor for democracy itself," Gore imagined this network of networks enabling widespread participation from citizens, its fora giving rise to "a new Athenian Age of democracy."[4]

If the Internet has been embraced by politicians, does it succeed in fulfilling the vision of its early-nineties boosters? Does it short-circuit the few-to-many discourse that dominated the twentieth century (in which politicians or mass media disseminate their message to a passive mass audience), replacing it with a many-to-many dialogue (in which both means of production and distribution of political ideas are available to all)?

In other words, does the Internet form a new public sphere or does it merely perpetuate the existing conditions? For philosopher Jürgen Habermas, who defined the term, *public sphere* refers to "a network for communicating information and points of view" in which democratic deliberation takes place.[5] By exchanging views on matters of common concern in a rational process of debate, citizens formed opinions that then shaped political decisions. Habermas understood there to be only one universal public sphere that all citizens would be able to take part in and saw the press as a critical check on political discourse.

But Habermas's account of the public sphere was a eulogy. The bourgeois public sphere that he observed emerged in coffeehouses and salons in the eighteenth century in which middle-class citizens, all (in theory) of equal status, discussed issues and reason prevailed. As capitalism developed, Habermas concluded, the uneven distribution of wealth and the emergence of mass media extinguished the capacity of citizens to have their voices heard, damaging the public sphere.[6]

If politics under the Internet is to do more than just perpetuate business-as-usual, it will have to structurally alter the political process itself. Advocates of the Internet as a venue for deliberative democracy and advocates of online democratic mobilization suggest that it can do just that.[7] Two examples that we could take as representing a turning point in the use of the Internet for political purposes highlight these different aspects of how democracy might be transformed online.

The first came in the aftermath of 9/11, taking the form of an Internet-based dialogue to address what should be built on the World Trade Center site. Sponsored by the Civic Alliance to Rebuild Downtown New York and the Port Authority of New York and New Jersey, the 2002 "Listening to the City Online" program brought eight hundred citizens into deliberative dialogue about how to redevelop the site and how to create a memorial for the victims. Through a process of structured, guided discussion and deliberation, participants contributed their positions to decision makers.[8]

Three years earlier, during the 1999 meeting of the World Trade Organization (WTO) in Seattle, online activists mobilized protests against globalization in what came to be known as the Battle in Seattle. Like the Internet itself, these actions were distributed, the product of many individuals and groups acting on their own initiative. Web sites like SeattleWTO.org and Seattle99.org linked together activists, helping incoming protestors find local hosts to stay with. Listservs allowed M2M discussions between activists. A fake Web site (gatt.org) parodied the WTO while the Independent Media Center (indymedia.org) served as a more legitimate media outlet for participatory journalists. This alternative news organization helped relatively powerless groups frame and disseminate their message as well as exercise leverage against a powerful, international organization.[9]

At a glance, in the two examples above, the Internet's promise to become a new democratic public sphere seems to be fulfilled. In "Listening to the City Online," deliberation processes took place online through a series of virtual meetings and dialogues. In the Battle in Seattle, creative uses of Internet media mobilized publics to real-world action and garnered widespread attention to their cause. Here we see two different yet overlapped modes of democratic action. The first mode is deliberation, referring to the involvement of citizens in decision making by engaging them in discussions on issues, soliciting their opinions on various points of view, and encouraging them to converse with one another to think critically about choices they make together. The second mode is mobilization, referring to the creation of broad social networks of people around a shared interest in blocking or promoting social change. Public deliberation can be a prelude to mobilization or a form of mobilization and can anticipate a more democratic future. Democratic mobilization elicits deliberation over goals, strategies, and tactics.

Still, priorities differ: deliberative democracy prioritizes the centrality of public talk in democratic governance; democratic mobilization emphasizes public activism against undemocratic forces.

While the examples seem convincing, is the Internet really a new, democratic public sphere in which those who rarely participate inform themselves, deliberate important issues, find their political voice, and thereby reshape political culture, if not public policy? Is the Internet likely to function as a democratic instrument for grassroots activists to challenge the authority and hegemony of powerful economic and political elites at home and abroad?

To be sure, some people successfully engage in online discussions that might eventually become deliberative processes, while others mobilize in collective movements. But in both cases there are inevitably more people who do not participate even though they are active Internet users. To complicate the issue, even if there are some successful examples of mobilization, not all projects are democratic; some are uncivil, anarchic, and even undemocratic.

The growing importance of the Internet does not mean that the medium necessarily fosters greater democracy. Skeptics point out that the kind of big media channels that dominated conventional politics in the pre-Internet era continue to dominate discourse in today's networked world.[10] Moreover, the vast amount of information available on the Internet is more than a storehouse of public knowledge; it is also a treasure trove for antidemocratic forces intent on monitoring, scrutinizing, and sanctioning dissidents in particular and citizens in general. Thus, for example, citing its need to cooperate with the laws of the countries it does business in, Yahoo! recently helped the Chinese government identify a journalist who was subsequently jailed for divulging state secrets.[11] So, too, other scholars suggest that the very vastness and nonhierarchical nature of the Internet makes finding authoritative information difficult or, conversely, that our ability to tailor information to our own interests means that we effectively put blinders on with regard to matters that *should* be of concern to us, even though they may not be within our narrow frame of reference.[12]

In this chapter, we do not attempt to find a definitive answer as to whether the Internet promotes democracy or if it is a new public sphere. More modestly, we argue that the Internet is a convivial milieu in which various political uses are thriving and new tools for political criticism and commentary are emerging. We show this by first comparing online efforts to promote deliberative democracy and democratic mobilization to understand how activists use the Internet to advance democracy. Beyond that, we look at blogging and remix, new types of political participation classifiable neither as mobilization nor as deliberation.

The Internet as Convivial Medium

To get a better grasp on the Internet's impact on politics, we turn to philosopher, educator, and social critic Ivan Illich's idea of a convivial society. Illich sought a postindustrial society that would maximize individual creativity, imagination, and energy rather than one that aimed to maximize outputs, as is the case in industrial societies. Behavior in a convivial society is composed of autonomous and creative interaction among individuals and their environments, a sharp contrast to the conditioned response of individuals living with reified social relations in an artificial, man-made milieu. Conviviality, for Illich, is by no means unfettered individuality. To the contrary, in his view, individual freedom is realized through interdependence and, as such, has an intrinsic ethical value. Ultimately, a convivial environment favors the freedom, autonomy, equality, and creative collaboration conducive to democracy.[13]

Popular in the 1970s, Illich's theories inspired some early pioneers of personal computing, such as Lee Felsenstein and Seymour Papert. Felsenstein founded the seminal Homebrew Computer Club (a Silicon Valley group that provided fertile ground for the development of the personal computer, its members including Steve Jobs and Steve Wozniak) and designed the first portable computer, the Osborne 1. Papert created the Logo programming language and was a proponent of using computers to educate children.[14] Although these individuals succeeded in creating convivial milieus in computer culture and in schools, these were still local conditions dependent on face-to-face interaction. Moreover, if the personal computer laid the groundwork for a convivial society, it was primarily a tool for individuals, difficult to extend to groups or to society as a whole. In contrast, since the Internet is dependent on the principle of connection it is inherently a single convivial milieu on a global scale.

The Internet is the product of the convergence of communication technologies. It is a network over which a variety of media can flow without regard to their specific qualities. Thus, the Internet can emulate traditional media such as print, radio broadcasting, telephony, television, and other existing technologies. But the Internet not only facilitates the traditional modes of one-to-one communications (as with telephone and telegraph) or one-to-many communications (as with newspapers and television), it also permits new forms of many-to-many and peer-to-peer communications and sharing.

Crucially, this convergence is achieved at low cost. The Internet is a relatively inexpensive technology, cheaper than even the telephone. It is inexpensive (even free) and easy to publish material on the Internet, either through the Web or by e-mail. Texts and images can be published without editorial interference and can rapidly achieve wide circulation.

Through Internet cafés and other public access points, the Internet is broadly available not only in developed countries but also in developing countries. Even for small organizations, the Internet already offers the least expensive means of communication capable of global reach.

The inherently decentralized character of the technology makes it relatively difficult to control or censor. While it is not nonhierarchical, the Internet is a network of networks, less hierarchical than previous media and communication technology. Although censorship, surveillance, and disruption can and does occur, it is limited by the nature of the network. A firewall can be set up and filtering can be applied, as China has done, but sufficiently savvy users usually can find ways to get messages to their intended destinations.[15] Moreover, the sheer volume of information flooding the Internet limits the effectiveness of most surveillance and censorship efforts.

Just like any other technology, the Internet can reinforce existing relationships between those who control technology and those who consume its products. Older Internet applications maintained a distinct separation between producers and consumers. For example, early Web sites functioned like bulletin boards or newspapers. Readers were meant to consume the content and had no tools with which to respond to, or change the content. But new applications and delivery platforms such as blogging software (for example, Blogger and Typepad), community-oriented content management systems (Drupal and Elgg), news-feed reading and aggregation software (My Yahoo!, Planet Aggregator, Feedreader) and video-distribution platforms (Google Video and YouTube) allow users to personalize their Internet intake and create their own content, giving rise to amateur producers who are at the same time also consumers, audiences, critics, and fans of—as well as collaborators with—other amateurs.

Characterized by convergence, low cost, broad availability, resistance to control, and the emergence of amateur production, the Internet is a convivial medium with a greater scope for freedom, autonomy, creativity, and collaboration than previous media. To be clear, however, there is nothing inherent in Internet technology that automatically achieves this potential. Unlike many theorists of postindustrialism, postmodernism, and information society, we do not see technology as a causal agent having a pivotal role in social change.[16] Nor do we see technology as neutral, its roles and outcomes completely determined by users. Rather, we understand artifacts as both constituted by society and constituting society. Social arrangements and contexts around the technology—human choices and politics—are key in deciding the impact of the Internet on politics, but the inherent limits and possibilities of technologies are also very important factors.

Online Deliberation

Theories of deliberative democracy attracted great attention in the 1990s when Jürgen Habermas and John Rawls published key treatises on the topic. In *Between Facts and Norms,* Habermas revisited possibilities for a renewed public sphere centered on inclusive, public deliberations free from inequalities and coercion. Deliberative citizens, under Habermas's model, would follow the force of the better argument, functioning "as a sounding board" for the political system.[17] Rawls's *Political Liberalism* begins with the assumption that people have different comprehensive views of the common good. The best way to build a stable society that respects these differences is, Rawls argues, to institute a deliberative democracy in which citizens have the knowledge and desire "to follow public reason and to realize its ideal in their political conduct." Rawls points out that deliberative citizens "explain to one another . . . how the principles and policies they advocate and vote for can be supported by the political values of public reason." They deliberate "as if they were legislators and ask themselves what statutes . . . they would think it most reasonable to enact."[18]

In the late 1990s, scholars and advocates of deliberative democracy turned to the Internet, envisioning cyberspace as a new, democratic public sphere in which P2P exchanges and M2M forums would enable large numbers of citizens to deliberate on a broad range of public issues and express their informed, thoughtful views in ways that would reflect and influence public opinion as well as urge, if not compel, cooperation by political decision makers.

Advocates of online deliberation understand it as not merely an online version of offline deliberation, but also as a way to solve problems associated with face-to-face discussions. If an ancient barrier to democracy was that only a few people could assemble in one place at one time to carry on public discourse, the Internet enables vast numbers of people to assemble in virtual space. As e-Democracy advocates Stephen Coleman and John Gøtze suggest in their defense of online deliberation, "the asynchronous nature of online engagement . . . makes manageable large-scale, many-to-many discussion and deliberation," overcoming the problem of getting people together to discuss issues at the same time.[19] Moreover, although no one has time to deliberate on every issue, the Internet can host an unlimited number of forums, and citizens can participate in issues of importance to them.

Online deliberation can bring together a mix of people who would not ordinarily encounter each other or talk to each other in everyday life. Online forums designed to ensure diversity may reduce common misunderstandings across class or racial divides, promote a degree of empathy, and foster greater mutual respect if not consensual agreement. By contrast, there is evidence that deliberation among like-minded people tends to produce greater polarization and extremism on public issues.[20]

The practice of online deliberation has, thus far, been modest. For the most part, online deliberation has made online forums available to citizens and non-profit organizations that invite the public to deliberate local issues and has offered online services to elected officials and political bureaucrats who want to consult with a broad range of citizens and stakeholders.

The most common model of online deliberation adapts offline deliberative practices to the Internet. James Fishkin's innovative *deliberative polling,* which combines face-to-face talk and public-opinion surveys, is now being conducted online, reducing organizing costs and participant inconveniences.[21] Similarly Beth Noveck recommends adapting the Citizens' Jury model to an online environment. Introduced in Great Britain in the 1990s, a Citizens' Jury consists of a randomly selected panel of citizens who act as representatives of their community, meeting for several days at a time to examine a public issue. The jury hears amateur and expert witnesses, deliberates on the issue, and presents recommendations to the public. Noveck's idea is to assemble Citizens' Juries online, employing new media tools to "delineate a problem, visualize and map out causes and effects, think through options, provide information, and collectively design solutions."[22]

Other developers set out to create forms of deliberation specific for the Internet, creating online forums to bring information, rationality, reciprocity, and civility to the Internet's new public sphere. Unchat, developed with the guidance of deliberative democracy advocate Benjamin Barber, promises to marry "the proven value of facilitated group conversation to the efficiency of the Internet to create productive, democratic decision making."[23] Web Lab hosts online dialogues "designed to avoid the pitfalls and weaknesses of typical computer bulletin-boards: the 'drive-by' postings encouraged by the Internet's easy anonymity and fluid boundaries; the assertion of polarized positions, where the give-and-take of civil discourse would have more social value; and the pandering to appetites for quick sensation rather than the creation of a real forum."[24] E-Liberate, developed by Evergreen State College, applies Robert's Rules of Order to online discussions.[25] Information Renaissance sponsors online forums that assemble "members of the public to learn about a complex issue and discuss it with subject experts, public advocates, and policy makers." Online participants access a briefing book, participate in dialogues, consult experts, and make recommendations.[26]

Yet other groups use the Internet to assist face-to-face forums. The America*Speaks*' 21st Century Town Meeting employs Internet technology to merge small, face-to-face group dialogues with large-scale gatherings, followed by online deliberation.[27] The Center for Wise Democracy favors Wisdom Councils, in-person deliberations extended with "group-ware."[28] Others promote Internet-assisted consultation and rulemaking to enable the citizens and stakeholders to give informed advice to public officials who make laws, policies, and administrative rules.[29] Big players such as NGOs and government bodies are interested as well: the Organisation for Economic Co-operation and Development (OECD) promotes citizen consultations, including online forums and bulletin boards, citizens' juries, and e-community tools.[30]

All three approaches emphasize rule-bound deliberation. This is typical of deliberative democracy: rules are intended to foster equality (everyone should be able to contribute), diversity (of participants and positions on issues), and common goals.[31] Procedural rules are important as well. For example, "The navigation of Unchat is expressly designed to promote . . . deliberation. A participant wanting to jump into a conversation must first pass through the library. . . . After the library, participants may be asked to take a quiz."[32] Most online forums have a facilitator "to provide discursive focus, stimulate groups into interacting constructively, build a sense of team spirit or community, referee, troubleshoot and keep time."[33] The facilitator might be a professional or participant whose job is to keep discussions on track and enforce rules of discourse.

Web Lab embeds rules in code. Participants must register, "creating a 'screen name' and password, providing an e-mail address, some basic information about themselves, and a short self-description." The software assigns a small number of diverse individuals to a dialogue group, which is then closed to new members but open to online discussions. Participants receive brief biographies of other members. Discussions are self-moderated but observed by a monitor to "watch for technical glitches, spot interesting dialogues to highlight in the Featured Posts section, or bring important issues you ask us to address to our attention."[34] Unchat software allows participants to take turns wielding the gavel to enforce a fairly strict set of rules; however, users may seek to modify the rules.[35] Finally, e-Liberate has embedded rules promulgated by online displays regarding what "legal actions" are available to participants at any point in a discussion.[36]

The degree to which online forums are preoccupied with rules, procedures, and moderators varies. But most groups that host online forums see rules as a matter of survival, a critical means of defense against Internet spammers, trolls, and ideologues who might seek to disrupt or polarize dialogue, rather than

Box 3.1 On Deliberation

From Berkman Center for Internet & Society, Harvard Law School, *Online Deliberative Discourse Research Project* (2000), http://cyber.law.harvard.edu/projects/deliberation/.

Most self-organizing online communities reflect a judgment we share: that democracy is the best means by which cyberspaces may be governed. The demonstrated nature of online communities as places where communication and discussion are valued suggests that deliberative discourse (i.e., reasoned communication that is focused and intended to culminate in group decision-making) is the form of democracy most prized online. Cyberspace also naturally supports another feature that is highly desirable for deliberative discourse: equality among the participants, including especially an equal ability to disseminate information to contribute to reasoned decision-making.

Governance of online communities requires the consent of the governed in a way and to a degree that physical communities do not. Coercive power over the body of a participant, the ultimate if often unspoken tool of offline governance, does not exist over the incorporeal citizens of online communities. Control by those in authority online ends, as does that of offline counterparts, at the borders of whatever spaces comprise the polity. However, unlike in offline jurisdictions, online authorities have no significant means by which to force their citizens to remain in those spaces. This is a difference at the most basic level: not even the presence of members of self-organizing online communities is assured. For any reason or no reason at all a member can simply leave the community, sacrificing whatever social investment he has made there, usually without financial or physical loss. This essential fact of online participation demands a structure that is encouraging, egalitarian, productive and rewarding. If the process of online discourse and decision-making is unpleasant, elitist, non-productive and/or time wasting, then people will vote with their browsers by failing to log on. . . .

The Internet is perceived as the next great leap forward in political and organizational interaction. However, the technology on which it rests is complex and often hidden from view. Computer programmers are in some respects the cyberspace equivalent of politicians' smoke-filled back rooms. If political processes are to move online, it is essential that the code, which facilitates and constrains the discussion and measures the community's opinion, must be as open and transparent as the systems of democratic government that we most admire. Public interest sponsorship of such code . . . is critical if online deliberation is to become a trusted and valuable tool of democracy.

participate in it.[37] And yet, those same rules create a high barrier of entry to dialogue, undoing the very ease of access that the Internet affords.

This preoccupation with rules is consistent with online deliberative democrats' modest aspirations for democracy. Effective online forums aim to produce three results. First, participants become more thoughtful and their views are taken more seriously. When deliberators know that their informed voice is being heard, they are likely to overcome their distrust of public officials.[38] Second, public officials become more trusting of informed citizens and, by listening to them, achieve greater legitimacy in their legislative and policy-making functions.[39] Third, online deliberation "deepens the relationship between decision makers and the public," inviting people to become more engaged in civic life while expanding "the scope, breadth, and depth of government consultations with citizens."[40] The ideal result is a partnership that, in the words of Coleman and Gøtze, "acknowledges a role for citizens in proposing policy options and shaping the policy dialogue—although the responsibility for the final decision or policy formulation rests with government."[41]

Organizers of online forums clearly desire to make deliberation safe for public officials and urge them to involve themselves in deliberation. They believe this will encourage greater public participation in discussions and increase the likelihood that public officials will heed the public's informed voice. From the vantage point of deepening democracy, this desire to include public officials is both promising and problematic.

It is promising in the sense that it encourages civic engagement and it potentially closes the gap between citizens and their representatives. This fosters a sense of efficacy, builds social capital, and encourages popular participation in public life. Furthermore, to the extent that deliberative forums deliver thoughtful recommendations, lawmakers and policy makers will have a greater incentive to solicit public advice and be guided by citizens in the future.

On the other hand, a partnership between citizens and public officials is problematic. When Benjamin Barber called for deliberative democracy in his 1984 book, *Strong Democracy,* he concluded that deliberative talk must be linked to citizen decision making and democratic activism. He argued that citizens are sovereign and have a right not only to deliberate but also to decide public issues and mobilize against dominant elites that monopolize decision-making power.[42] By contrast, online deliberation is not premised on citizen sovereignty, decision-making authority, or political struggle against dominant elites. Rather, it emphasizes constrained talk and mostly accepts the current distribution of power by ceding decision-making authority to public officials who—partners or not—rarely defy the interests of dominant elites.

To be sure, online deliberative democrats help fulfill the convivial potential of the Internet. They provide many people access to forums for deliberation on a range of public issues. They try to involve decision makers in online forums, thereby assuring participants that their voices will be heard. They seek to build a new public sphere in which rationality rules, citizen's voices are heard, and public officials heed the *demos.*

In theory, the growth of this new public sphere should result in greater citizen satisfaction, greater government legitimacy, and greater political stability within established governmental jurisdictions—such as cities, states, and nations. Online deliberative democracy does not directly address ongoing inequalities that threaten individual liberty, autonomy, creativity, and democratic collaboration. Nor does it directly address issues that reach beyond established government jurisdictions to the global arena. In effect, it contributes to democratic government where it more or less exists, but it cannot contribute to struggles to contest the influence of local and global elites who use economic, as well as political, power to undermine human rights, perpetuate injustices, and defeat democratization.

Online Mobilization

In cases where rational dialogue does not seem possible, online mobilization offers an alternative. Critics of postmodernism, such as Alain Touraine, point out that even in present-day democracies, the state, the market, and the media are gradually diminishing the liberty of the individual, failing to guarantee freedom, equality, and fraternity.[43] In response, Habermas suggests that new social movements outside of the traditional public sphere are developing.[44] These new social movements are broad alliances of people sharing an interest in blocking or promoting social change, for example, movements against globalization (since most governments endorse globalization, it tends not to inspire a social movement to promote it), for or against immigration, for or against abortion, or for or against various human rights such as the rights of homosexuals to marry, gender equality, and so on.

Historically, activists have been quick to incorporate media such as publications, radio, television, and film to mobilize their constituencies to action. During the past decade, progressive activists have turned to online forms of communication, community building, and resistance.[45] The recent upsurge of online mobilization includes global support for peace movements, opposition to the Iraq war, and protests against neoliberal organizations.[46] Online mobilizations have also developed at the local and national levels, yet involve actors focused on global issues. Some of the most prominent examples include online activism in support of the Zapatistas in Chiapas, Mexico, the Free Burma Coalition, and the pro-democracy movement and political revolution in Indonesia in May 1998.[47]

Usually, mainstream media, with or without collaboration by activists, have played an important role in portraying political activism. Although this could be beneficial in increasing exposure, it could also distort or simplify the messages that activists intended. The Internet allows activists themselves to frame their issues and shape their public identities.

The Internet allows online organizers to combine the advantages of one-to-one communication, one-to-many broadcasts, and many-to-many media. This enhances opportunities for activists to mobilize and promote their causes. Successful online actions, such as the worldwide antiwar protest initiated by MoveOn.org, demonstrate that the Internet can facilitate global activism more directly and quickly than previous technologies.[48] The Internet's broad availability, along with its one-to-many and many-to-many modes of communication, make it possible for an organization to quickly and affordably reach a large group of people while targeting communications to specific parties. Mobilizing online also enables activists to talk back, responding by e-mail or through platforms that allow for questions and elaborations. The result is a partial move from face-to-face to faceless tactics, with protest happening online in coordinated (yet physically separated) actions around the world.

For online mobilization at the local and national levels, the Internet provides a global dimension. The Zapatista movement in Chiapas is an example. One analysis of the communication dimension of the movement observed that the "most striking thing about the sequence of events set in motion on January 1, 1994, has been the speed with which news of the struggle circulated and the rapidity of the mobilization of support which resulted."[49] The Internet and the networks of the Association for Progressive Communications enabled the Zapatistas to bypass government control and get out their message. Global communication networks facilitated support activities and organized protests in more than forty countries, from marches, raves, and readings in San Francisco to a rally in the Piazza del Popolo in Rome.[50]

Another example is the case of "Free Burma," in which Burmese dissidents used the Internet to create a global network of resistance against the military junta in power. Started by a Burmese student living in exile, the network enabled exiles from Burma who shared similar political concerns to coordinate, bringing issues such as human rights violations in the country to the media attention, and to put pressure on the military regime by encouraging companies to stop investing or operating in Burma. The Internet helped various organizations to coordinate their activities, allowing them to orchestrate ground actions as a collective instead of as a set of disparate individuals.[51]

Online organizing tools have the potential to increase the scale of organizing efforts while keeping costs low. With e-mail, it is possible to send out one million announcements, donation solicitations, and calls for action for next to nothing. E-mail functions not only as a one-to-many form of distribution but also as a P2P form of distribution as individuals forward messages to their like-minded friends. Indeed, as the Seattle protests against the World Trade Organization demonstrated, e-mail can be very effective in the preparation for and the follow-up to demonstrations.[52]

Nevertheless, the Internet is rarely the sole theater of activity for social movements. The success of mass events like the Seattle protests requires the use of multiple media and organizing tactics. Intermodality between the Internet and other media networks, as well as between cyberspace and geographical place, is generally necessary to allow activists to produce and disseminate information as well as to organize and mobilize for action.[53] During the successful pro-democracy movement in Indonesia in 1998, links between the Internet and more traditional media and existing social networks were crucial. After the Mexican Army countered the guerilla tactics of the Zapatista Army of National Liberation, the Zapatistas turned to guerilla radio and the Internet to get their message out. Similarly, the hybrid use of text messages sent by mobile phone and messages sent by e-mail in the EDSA II "People Power" movement in the Philippines from 2000 to 2001 is another example of the strategic importance of intermodality.[54]

The Internet is not a passive medium; on the contrary, it challenges conventional structures for organizing social movements, encouraging the undoing of hierarchical and centralized communications in favor of more decentralized and distributed organizational structures.[55] Previous communication technologies, even grassroots organizing techniques such as phone and fax trees, required somewhat hierarchical structures. By contrast, (unmoderated) mailing lists and P2P applications such as e-mail make it possible for activists to organize quickly, with little logistical coordination or organizational oversight.

Thus, the Internet's inherent conviviality enables rapid, widespread mobilization. Although it is faster to mobilize participants around shared issues, such bonds are not necessarily long lasting. Online protest groups tend to be single-issue based, ephemeral, and shortsighted in terms of the scope of change they wish to effect.

Although it possesses some characteristics that favor activist movements, the Internet is not a tool that can resolve all problems intrinsic to democratic mobilization. In fact, the Internet has the potential to amplify movements of any kind, regardless of their ideologies, purposes, and goals. Online mobilization is not inherently democratic by any means. Anarchic, radical fundamentalist, and

Box 3.2 On Mobilization

From Geertz Lovink and Florian Schneider, "A Virtual World is Possible: From Tactical Media to Digital Multitudes," (2003), http://www.makeworlds.org/node/22.

By the end of the nineties the post-modern 'time without movements' had come to an end. The organized discontent against neo-liberalism, global warming policies, labour exploitation and numerous other issues converged. Equipped with networks and arguments, backed up by decades of research, a hybrid movement gained momentum, wrongly labelled by mainstream media as 'anti-globalisation.' It seemed one of the specific flags of that movement, that it hasn't been able and willing to answer the question, which constitutes any kind of movement on the rise, any generation on the move: what's to be done? There was and there is no answer, no alternative 'either strategic or tactical' to the existing world order, to the dominant mode of globalisation.

And maybe this is the most important, and liberating, conclusion: there's no way back to the twentieth century, the protective nation state and the gruesome tragedies of the 'left.' It had been good to remember, but equally good to throw off, the past. The question 'what's to be done' should not be read as an attempt to re-introduce Leninist principles in whatever form. The issues of strategy, organization and democracy belong to all times. We neither want to bring back old policies through the backdoor, nor do we think that this urgent question can be dismissed with the (justified) argument of crimes committed under the banner of Lenin. When he looks in the mirror Slavoj Žižek may see Father Lenin, but that's not the case for everyone. It is possible to wake up from the nightmare of historical communism and (still) pose the question: what's to be done? Can a 'multitude' of interests and backgrounds ask that question, or is the agenda the one defined by the summit calendar of world leaders and the business elite?

Nevertheless, the movement has been growing rapidly. At first sight, by using a pretty boring and very traditional medium: the mass-mobilization of tens of thousands in the streets of Seattle, hundreds of thousands in the streets of Genoa. Tactical media networks played an important role in its coming into being. From now on pluriformity of issues and identities was a given reality. Difference is here to stay and no longer needs to legitimize itself against higher authorities such as the Party, the Union or the Media. This is the biggest gain compared to previous decades. The 'multitudes' are not a dream or some theoretical construct but a reality.

If there is a strategy, it's not contradiction, but complementary existence. Despite theoretical deliberations, there is no contradiction between the street and cyberspace. The one fuels the other. Protests against WTO, neo-liberal EU policies, and party conventions are all staged in front of the gathered world press. Indymedia crops up as a parasite of the mainstream media. Instead of having to beg for attention, protests place under the eyes of the world media during summits of politicians and business leaders, seeking direct confrontation. Alternatively, symbolic sites are chosen such as border regions (East-West Europe, USA-Mexico) or refugee detention centres (Frankfurt airport, the centralized Eurocop database in Strasbourg, the Woomera detention centre in the Australian desert). The global entitlement of the movement adds a new layer of globalisation from below to the ruling mode of globalisation, rather than just objecting to it.

even terrorist groups also employ online mobilization as part of their struggles and strategies. Extremist groups such as Al-Qaeda, as well as smaller radical fundamentalist groups such as Stormfront in the United States and Laskar Jihad in Indonesia, have used the Internet to mobilize. Just like advocates of democracy, extreme fundamentalist groups rely on the Internet to widen their scope of operation, reach broad audiences, and mobilize to gain more influence and power.

Backing into the Future?

In practice, online mobilization has been more successful than online deliberation. In contrast to online deliberation's rule-bound systems, online mobilization's looser, distributed nature is more in keeping with the Internet's informal, convivial nature and is thus able to thrive online.

It is striking, however, that both online deliberation and online mobilization rely so much on traditional tactics. But as Marshall McLuhan suggests, this is the norm rather than the exception: "When faced with a totally new situation," he says, "we tend always to attach ourselves to the objects, to the flavor of the most recent past. We look at the present through a rearview mirror. We march backwards into the future." In short, online activism scholar Graham Meikle describes this as "backing into the future."[56] Most online deliberation projects move offline forums to cyberspace, connect to offline forums, or emulate offline forums. The tactics of online democratic mobilization, such as online petitions and virtual sit-ins, are derived from traditional activities, such as paper petitions and actual sit-ins.

But backing into the future does not mean that online activists are not being innovative. Rather, it suggests that they frame online mobilization by the sociotechnical ecology of traditional mobilization. In effect, online activists reinvent familiar activist methods. Indeed, the success of online mobilization may be related to its familiarity.

That activists are backing into the future does not prevent online innovations—such as site hijackings, hacktivism, e-mail distribution trees, smart mobs or flash mobs from emerging. While these phenomena can be seen as digital analogues of traditional tactics such as sabotage, letter writing, phone and fax trees, and street demonstrations, they have qualities that make them unique and provide a foundation for further innovation.

Although online deliberative democracy and online democratic mobilization are central in academic discourse on online politics, they represent only a fraction of online political activities. The most vibrant political activities in network culture are not actually located in collective political actions such as deliberation and mobilization, but rather are located between private and private, between private and public, and between publics. They emerge in the overlapping domains of politics and culture, simultaneously among multiple layers of social networks, between multiple networks of individuals, and between individuals and collectives, creating a sphere of networked politics. Popular examples of such activities include online political art, cartoons, and videos. In this chapter, however, we will focus on blogging and remix as they have the lowest threshold of skill and technology necessary for entry and are currently the most common of these activities.

Political Blogging

Defined by Wikipedia as "Web-based publication[s] consisting primarily of periodic articles (normally in reverse chronological order),"[57] blogs allow their creators to frequently and easily update information and to elicit discussion among their readers by recording comments.

In politics, blogs became popular during the 2004 United States Presidential campaign. Until then, even the most popular blogs received only a tiny proportion of the Web traffic that major media outlets attracted, and politicians did not see them as capable of a serious political role.[58] The turning point was Howard Dean's *Blog for America,* which showed how a blog could be used for building social networks of political support. Dean's employment of blogging and the rapid rise in the popularity and proliferation of political blogs that year demonstrated their potential to politicians. In subsequent years, blogging has become much more popular, with about fifty-seven million American adults reading blogs by 2006.[59]

Admittedly, political blogs are only a fraction of the blogosphere. All sorts of content can be found on blogs—from announcements of new gadgets to discussions of films and television to firsthand accounts of child-rearing—but many of these, such as Boing Boing, one the most popular blogs in the world, blur this distinction by also commenting on political matters.[60] Moreover, many political blogs have devoted readerships. The progressive American blog Daily Kos, for example, attracts about six hundred thousand visitors per day and has between fourteen and twenty-four million visits per month, making it one of the most popular collaborative blogs in the world.[61] In contrast, the *Nation,* which describes itself as "the most widely-read weekly political opinion magazine in America" had only 187,000 subscribers in 2005.[62]

Some observers see blogs as a catalyst for change, a people's media and an empowering tool. They see the rise of blogs beginning an era of citizen journalism in which the marginalized can play a greater role in making, rather than merely consuming, news. Others argue that public debate would be dramatically revitalized if politicians would all start blogging.[63]

But what role do blogs really play in empowering society? Are they a real breakthrough in online politics? We identify several problems in the current practice of political blogging. First, the blogosphere suffers from an unequal distribution of readers. While there are over a million bloggers in the United States posting approximately 275,000 new items daily, the average blogger has almost no influence on other blogs as measured by traffic. The distribution of links and traffic is skewed so that only a handful of bloggers get most of the readers. Generally speaking, these are either those who got established in the blogosophere early, when there was little content, or were well-known journalists or politicians such as Ariana Huffington.[64]

This tendency shows that the blogosphere is not an exemplary public sphere in which everybody's voice is heard. If this could be also seen as a selection process, weeding out the "bad" blogs, it also favors players who got in early or who make outlandish statements to attract readers.[65]

Second, some studies show that rather than creating a new public (blogo) sphere, bloggers tend to be polarized along ideological lines. Lada Adamic and Natalie Glance's study on the American political blogosphere finds "liberals and conservatives linking primarily within their separate communities, with far fewer cross-links exchanged between them. This division extended into their discussions, with liberal and conservative blogs focusing on different news articles, topics, and political figures."[66]

But such studies give rise to questions. Does the polarization of the American blogosphere mirror society itself? Or does the blogosphere cause this polarization? Is the polarization it causes substantially greater than through other media?

These are not easy to answer. We can hypothesize that the culture of linking in the blogosphere may create more exposure to divergent ideas than people otherwise experience in real space; thus, we could suggest that it is not a contributing cause of existing political polarization. Or, to the contrary, we could argue that the vast body of metadata produced by tagging content on services like Technorati makes it possible for a blogger to easily find information that confirms what she or he already believes, reinforcing polarization. Nevertheless, that same social metadata would also increase serendipitous exposure to information that the blogger might disagree with, producing a different result. Empirical research is needed to answer such questions.

If polarization is one potential problem, another is that there is no central organization to the blogosphere and little consensus among bloggers with regard to many key issues. This creates a virtual Tower of Babel in which voices tend to become so particular and so exclusionary of other views as to be unable to communicate to each other or to a broader audience. On the other hand, the proliferation of unique points of view in the blogosphere may encourage genuinely individual voices to emerge and perhaps even foster real dialogues (as opposed to the watered-down positions distributed in mass media).[67]

The amateur status of bloggers also raises questions. Most bloggers are part-timers for whom blogging is a voluntary endeavor. Amateur bloggers do not, in general, have the resources and capacity to investigate material prior to publishing it. Thus, from a journalistic point of view, the credibility of blog entries generally cannot meet that of articles in mainstream media. On the other hand, the voluntary nature of blogging means that it is also a positive way for regular people to voice their opinions without going through the filtering effects of traditional journalism.

Yet another concern is the increasing tendency for the top blogs to resemble old media. Techmeme's top one hundred Leaderboard, for example, includes old media players such as the *New York Times,* the *Wall Street Journal,* and the *Associated Press.*[68] New blogs that manage to get into the top one hundred, such as TechCrunch, GigaOm, and Engadget, are far from the "people's media," heavily backed up by professional writers, editors, graphic designers, and marketing people. This shows that the blogosphere, too, is dominated by an existing structure of media power and ownership where individuals hardly have much space and power to play a significant role.

The blogosphere is not, and will never be, an ideal political sphere. Nor will it produce a common ground of rational communicative discourse. Nevertheless, political blogging is a unique online practice that expands the political sphere from the elites to commoners more effectively than previous Internet applications such as Web sites could. Moreover, while blogs might not be true examples of deliberative democracy, the kind of two-way communications that blogs facilitate between bloggers and people who leave comments on their blogs are facilitated with ease in the blogosphere. Although this will not create an ideal Habermasian public sphere, the multiple networked political spheres it generates are positive.

Political Remix

A variant of online activism takes place in the hybrid realm of culture and politics. The emergence of DIY audio- and video-authoring tools and sites to which individuals can easily upload the content they generate has fostered the rise of a remix, mash-up culture focused on politics and political issues. In the music industry, remix refers to alternative versions of audio or visual compositions derived from the original material. During the last few years, virally distributed remix videos and ads with political messages have become quite popular.

Political remix is not new; rather, it borrows from many movements within modernism and postmodernism—such as appropriation, collage, assemblage, Dada, surrealism, situationism, and punk rock—that alter images so as to subvert them, as well as audio movements—such as reggae, hip hop, and DJ culture—that do the same for music. As these earlier movements did, rather than supporting the status quo, remix changes visual imagery to convey a radical or oppositional message.

What contemporary remix offers is the ability to use digital technology to create convincing works that may not seem like remixes and then to distribute them on the Internet freely, widely, and in a reasonably short time. In the past, DIY culture was samizdat, distributed to a small group only, generally through mail or in localized communities. Although convivial, it could not reach beyond its narrow community.

As author William Gibson suggests, remix is the very nature of today's digital world.[69] New paths of information exchange between people keep growing, making the Internet a densely networked social milieu. This stimulates people to produce (and consume) by drawing information from multiple sources, remixing and making it into their own, and sharing it with others. The emergence of the social Web, or Web 2.0, enables a kind of "collaborative remixability," a phrase coined by Barb Dydwad to refer to "a transformative process in which the information and media we've organized and shared can be recombined and built on to create new forms, concepts, ideas, mashups, and services."[70]

Political remix engages mainstream political artifacts. Remix artist-activists recognize that the products of mainstream politics (such as political news on CNN) are source material that can capture widespread attention. By mashing up, remixing, or playing out alternative narratives, remix activists transform mainstream artifacts to promote new political messages. Many remix videos edit existing ads or news footage to create parodies and satires with new political meanings.

One well-known example of remix is the video *Bushwacked2.* Through careful editing of George W. Bush's 2003 State of the Union address, British satirist Chris Morris altered Bush's speech so that he would make pronouncements such as "We are building a culture to encourage international terrorism" and "I have a message to the people of Iraq: Go home and die."[71]

Hummertruth, a spoof on a Hummer H2 commercial, is another prominent remix.[72] By adding subtitles, social activist Jonathan McIntosh transformed a Hummer ad into a powerful commentary, suggesting that the vehicle was an icon of environmental degradation.

Box 3.3 On Blogging

From Ross Ferguson and Milica Howell, *Political Blogs: Craze or Convention?* (2004), http://www.hansardsociety.org.uk/files/folders/472/download.aspx.

The main political value of blogging is not to be found in politicians presenting themselves to an audience of potential voters, but in the dense networks of intellectual and symbolic intercourse involving millions of private-public bloggers. The blogosphere is characterised by three democratising characteristics. Firstly, it provides a bridge between the private, subjective sphere of self-expression and the socially-fragile civic sphere in which publics can form and act. As democracy becomes more sensitive to affective dimensions, attention is paid to a revalued recognition of subjective and intersubjective articulations. As several commentators have observed, it is often within the safety of private or familiar environments that people feel most able to speak as citizens. By allowing people to both interact with others and remain as individuals, blogs provide an important escape route from the "if you don't come to the meeting, you can't have anything to say" mentality.

Secondly, blogs allow people—indeed, expect them—to express incomplete thoughts. This terrain of intellectual evolution, vulnerability and search for confirmation or refutation from wider sources is in marked contrast to the crude certainties that dominate so much of political discourse. As Mortensen and Walker have explained: "We post to our blogs as ideas come to us. Daily, hourly, weekly; The frequency varies, but it is a writing that happens in bits and pieces, not in the long hours of thought that suit the clichéd image of the secluded scholar in the ivory tower. In this sense blogs are suited to the short attention span of our time that worries so many traditionalists. Blogs are interstitial for the writer as for the reader."

Thirdly, blogs lower the threshold of entry to the global debate for traditionally unheard or marginalised voices, particularly from poorer parts of the world which are too often represented by others, without being given a chance to present their own accounts. Blogs such as Hossein Derakhshan's Editor: Myself (http://hoder.com/weblog/), the South Korean Ohmy News (http://www.ohmynews.com/) and Blog Africa (http://blogafrica.com/) are refreshing additions to a global debate in which contributors have tended to be better at speaking for than listening to the world's least privileged.

It is as channels of honest self-presentation that blogs make their greatest contribution to democracy. If Walter Cronkite's famous sign-off, "That's the way it is" was the dictum of the world of media-represented factual certainties, "That's the way I am" is the dictum of a self-expressive culture where truth emerges in fragmented, subjective, incomplete and contestable ways.

Box 3.4 On Remix

From Eduardo Navas, "Turbulence: Remixes + Bonus Beats, 3 x 3: New Media Fix(es) on Turbulence," Jo-Anne Green and Helen Thorington, eds., Turbulence.org (2006), http://transition.turbulence.org/texts/nmf/Navas_EN.html.

Generally speaking, remix culture can be defined as the global activity consisting of the creative and efficient exchange of information made possible by digital technologies that is supported by the practice of cut/copy and paste.[a] The concept of Remix often referenced in popular culture derives from the model of music remixes which were produced around the late 1960s and early 1970s in New York City with roots in Jamaican music.[b] Today, Remix (the activity of taking samples from pre-existing materials to combine them into new forms according to personal taste) has been extended to other areas of culture, including the visual arts; it plays a vital role in mass communication, especially on the Internet.

Loops are essential to computer technology, for what else does the computer do but execute loops to know what it should be doing at all times?[c] In the days before the first computers, people did calculations manually, but at one point the need to have repetitive computations performed in a more efficient way became a concrete idea. And in 1945, with ENIAC, computers started to take over the role of human computers.[d] The concept of loops played a crucial role in culture at this time, as Pierre Schaeffer and Stockhausen were creating compositions consisting of loops that were performed not by humans but machines.[e] The loop in music became crucial for DJ culture, and DJ culture would meet digital culture in new media. This merging is crucial to Remix.

Remix is always allegorical following the postmodern theories of Craig Owens, who argues that in postmodernism a deconstruction, a transparent awareness of the history and politics behind the object of art is always made present[f] . . . Meaning that the object of contemplation . . . depends on recognition (reading) of a pre-existing text (or cultural code). The audience is always expected to see within the work of art its history. . . . Postmodernism [is], in effect, remixed modernism. . . . [H]istories are constantly revised in Remix.

But, to be clear—no matter what—the remix will always rely on the authority of the original song. The remix is in the end a re-mix—that is a rearrangement of something already recognizable; it functions at a second level: a meta-level. This implies that the originality of the remix is non-existent, therefore it must acknowledge its source of validation self-reflexively (even when it is a selective remix). In brief, the remix when extended as a cultural practice is a second mix of something pre-existent; the material that is mixed for a second time must be recognized, otherwise it could be misunderstood as something new, and it would become plagiarism. Without a history, the remix cannot be *Remix.*[g]

---

a. This is actually my own definition extending Lawrence Lessig's definition of Remix Culture based on the activity of "Rip, Mix and Burn." Lessig is concerned with copy-

4. Gore, "Remarks to the International Telecommunications Union," (speech, International Telecommunications Union, Buenos Aires, October 13, 1994), http://clinton1.nara.gov/White_House/EOP/OVP/html/telunion.html.

5. Habermas, *Between Facts and Norms: Contributions to a Discourse Theory of Law and Democracy* (Cambridge, MA: MIT Press, 1996), 360.

6. Habermas, *The Structural Transformation of the Public Sphere: An Inquiry into a Category of Bourgeois Society,* (Cambridge, MA: MIT Press, 1989), 33, 36, 41, 51, 171, 212.

7. For deliberative democracy on the Internet, see Antje Glimmer, "Deliberative Democracy, the Public Sphere, and the Internet," *Philosophy & Social Criticism* 27, no. 4 (2001): 21–39; For democratic mobilization in the wake of the Howard Dean campaign, which utilized online community-building resources, see Steven Levy, "Dean's Net Effect is Just the Start," *Newsweek,* March 29, 2004, http://www.msnbc.msn.com/id/4563310/site/newsweek/.

8. Civic Alliance to Rebuild Downtown New York, *Listening to the City Report of Proceedings,* 2002, http://www.weblab.org/ltc/LTC_Report.pdf.

9. Matthew Eagleton-Pierce, "The Internet and the Seattle WTO Protests," *Peace Review* 13, no. 3 (2001): 331–337.

10. Clay Shirky, "Power Laws, Weblogs, and Inequality," first published February 8, 2003, on the "Networks, Economics, and Culture" mailing list, http://www.shirky.com/writings/powerlaw_weblog.html.

11. "Yahoo 'Helped Jail China Writer,'" *BBC News,* September 7, 2005, http://news.bbc.co.uk/2/hi/asia-pacific/4221538.stm.

12. Cass R. Sunstein, "The Daily Me," in *Republic.com* (Princeton: Princeton University Press, 2001), 3–22.

13. Illich, *Tools for Conviviality* (New York: Harper & Row, 1973). The first use of the term *conviviality* to describe the sociotechnical landscape of the Internet is found in Merlyna Lim, "The Internet, Social Network and Reform in Indonesia" in *Contesting Media Power: Alternative Media in A Networked World,* ed. Nick Couldry and James Curran (Lanham: Rowan & Littlefield, 2003), 274.

14. Felsenstein, "Convivial Cybernetic Devices," interview with Kim Crosby, *The Analytical Engine 3,* http://www.opencollector.org/history/homebrew/engv3n1.html; Papert, *The Children's Machine: Rethinking School in the Age of the Computer* (New York: Basic Books, 1993), 141–144.

15. Richard Clayton, Steven J. Murdoch, and Robert N. M. Watson, "Ignoring the Great Firewall of China," (lecture, 6th Workshop on Privacy Enhancing Technologies, Robinson College Cambridge, UK, June 28, 2006), http://www.cl.cam.ac.uk/~rnc1/ignoring.pdf.

16. Daniel Bell, *The Coming of Post-Industrial Society* (New York: Basic Books, 1973), sees technology a central organizing factor in social transformation. For Jean François Lyotard, *The Postmodern Condition: A Report on Knowledge* (Minneapolis: University of Minnesota Press, 1984), and Jean Baudrillard, *Simulations* (New York: Semiotext(e),

right issues; my definition of Remix is concerned with aesthetics and its role in political economy. See Lawrence Lessig, "Free," *The Future of Ideas* (New York: Vintage, 2001), 12–15.

b. For some good accounts of DJ Culture see Ulf Poschardt, DJ Culture (London: Quartet Books, 1995); Bill Brewster and Frank Broughton, Last Night a DJ Saved my Life (New York: Grove Press, 1999); Javier Bláquez and Omar Morera, eds., Loops: una historia de la música electrónica (Barcelona: Reservoir Books, 2002).

c. Scott McCartney, "The Ancestors," Eniac, (New York: Walker and Company, 1999), 9–27.

d. Women working in the basement of the University of Pennsylvania's Moore School during WWII were called "computers" because they calculated (computed) ballistic missiles tables all day. See, McCartney, 95–97.

e. Rob Young, "Pioneers," Modulations (New York: Caipirinha, 2000), 10–20.

f. Craig Owens, "The Allegorical Impulse: Towards a Theory of Postmodernism," eds., Brian Wallis and Marcia Tucker, Art After Modernism (New York: Godine, 1984), 223.

g. DJ producers who sampled during the eighties found themselves having to acknowledge History by complying with the law; see the landmark law-suit against Biz Markie in Brewster, 246.

While most remix work concentrates on American politics, creative political artifacts in a non-American context exist as well, such as *Zendani Siasi,* a political music video with sequences that emphasize the Iranian regime's oppressive nature, and the well-known *French Democracy,* a machinima video that provides an alternative narrative on recent riots in France.[73]

Admittedly, it is possible to see remix productions as products of an apolitical youth culture, aestheticizing political issues. Still, these amateur productions exemplify how individuals can become actively engaged in the networked political sphere. Instead of blindly consuming political information, they express political views by producing and distributing their own works. While the responses do not always aim to mobilize opinions or lead toward tangible actions, political remix itself is ultimately a form of mobilization. Even when remixes do not endorse particular political views, remixing itself is a practice that inherently mobilizes resistance against top-down, mass-media messages. We agree with Henry Jenkins that the very ability of amateurs to express and disseminate their cultural preferences is an important aspect of democracy in contemporary society.[74]

Again, these works may not always foster democratic values. Participatory remix culture is not inherently democratic; it is convivial. It enables amateur producers to make statements that widen the spectrum of contestations over political meanings and practices. By opening a new avenue for participation, political remix culture potentially contributes to the formation of a more open, diverse, and egalitarian political segment in the networked publics.

Conclusion

Analyzing various modes of political participation, this chapter suggests that the Internet is not an ideal public sphere in which effective and robust public participation takes place. But this does not mean that political spheres generated by the Internet do not contribute to the democratic enhancement of a political system. The Internet is a political artifact that is politically constituted as well as constituting. A convivial medium, it is open to various uses.

We have argued that the Internet does provide a sympathetic milieu for deliberative democracy and democratic mobilization. At the same time, we also suggest that it is misleading to claim that online deliberation and online mobilization practices have really deepened democracy.

For the foreseeable future, online deliberation and online mobilization are forms of democratic participation that have different, sometimes conflicting, purposes. Rule-bound deliberation is slow and ponderous, emphasizes the acquisition of knowledge and expertise, focuses on government laws and policies, and succeeds when citizens partner with government officials in the service of good decisions, political legitimacy, and social stability. Democratic talk potentially deepens democracy where it more or less exists. In contrast, mobilization often requires quick, decisive action, emphasizes people's identities as historical agents of change, focuses on corporate influence within and beyond political jurisdictions, and succeeds when activists disrupt and disable undemocratic corporate entities and dictatorships from committing injustices. Democratic mobilization deepens democracy where it does not prevail. But, as the Internet speeds up the process and widens up the scale, online mobilization is always in danger of being too fast, too thin, and too many.

If the Internet will extend the reach of both deliberative democracy and democratic mobilization, it does so by backing into the future. Offline deliberations are either the explicit source of, or an implicit model for, developing online forums. While the Internet provides several advantages, such as the ability to host conversations with thousands of diverse participants at a time, face-to-face discussions also have advantages—especially where interpersonal trust is crucial for developing a consensus. Similarly, online mobilization has advantages that cannot be reproduced offline, but face-to-face gatherings may be necessary to sustain, organize, and focus political movements over time.

Whether it is a matter of democratic talk or action, then, we can expect to see hybrid forms of online and offline participation in the future.

Perhaps the more interesting question is whether these hybrids will be sufficiently creative, engaging, and energizing to motivate the apathetic, ambivalent, or immobile among us to give democratic participation a try. In general, activism has always required great self-sacrifice and a substantial time commitment, a price too high for most people. In contrast, the Internet opens the door to part-time deliberation and part-time activism.

Still, as we argue, the more promising forms of online politics are not bound within a framework of conventional politics. Activities that don't fit into the traditional political framework, such as political blogging and political remix, thrive on the Internet.

While it is not an ideal public sphere, political blogging creates an accessible medium for Internet users to communicate with other users. Despite its limitations and problems, we think that political blogging has the potential to give rise to new political positions and to bring together people around those interests.

So, too, the online participatory culture of remix facilitated by affordable digital technology, networked tools, and social software promotes a sense of cultural agency and fosters P2P networks, indicating that the Internet may become a more powerful gateway for people formerly on the sidelines to become local, even global, activists.

We want to emphasize that the Internet enables multiple, overlapped, and diverse networked political spheres to emerge. These are contested spheres that are sometimes messy, chaotic, segmented, and even anarchic. Not all o these aim to advance and deepen democracy, but within these convivial sphere individuals and groups have a greater ability to be political.

Ivan Illich says: "What are needed are new networks, readily available to th public and designed to spread equal opportunity for learning and teaching.' Little by little, we may be getting there.

Notes

1. R. Michael Alvarez and Thad E. Hall, *Point, Click, and Vote: The Future of Int Voting* (Washington DC: The Brookings Institution, 2004), 54.

2. Luke O'Brien, "YouTube and CNN Discuss 'Most Democratic' Presidentia bate Ever," *Wired Blog Network,* http://blog.wired.com/27bstroke6/2007/06/you and_cnn.html.

3. Kapor, "Where is the Digital Highway Really Heading?" *Wired* 1.03 (1993), www.wired.com/wired/archive/1.03/kapor.on.nii_pr.html.

1983), information technology is a central agent in the development of the postmodern condition. Similarly, Manuel Castells argues in *The Rise of the Network Society* (Oxford: Blackwell, 2000) that information technology is the principal driving force in the "network society."

17. Habermas, *Between Facts and Norms: Contributions to a Discourse Theory of Law and Democracy* (Cambridge, MA: MIT Press, 1996), 305–306, 359.

18. Rawls, *Political Liberalism* (New York: Columbia University Press, 2005), 217, 444–445, 448, 481.

19. Coleman and Gøtze, "Bowling Together: Online Public Engagement in Policy Deliberation," (report, Hansard Society, 2001), 17, http://bowlingtogether.net/.

20. James S. Fishkin, comments at the Networked Publics Conference and Media Festival, Los Angeles, CA, April 28–29, 2006; Cass R. Sunstein, "Deliberation Day and Political Extremism," The University of Chicago Law School Faculty Blog, comment posted on February 2, 2006, http://uchicagolaw.typepad.com/faculty/2006/02/deliberation_da.html. See also Cass R. Sunstein, *Republic.com.*

21. Shanto Iyengar, Robert C. Luskin, and James S. Fishkin, "Deliberative Preferences in the Presidential Nomination Campaign: Evidence from an Online Deliberative Poll," (research paper, The Center for Deliberative Democracy, 2005), 4, http://cdd.stanford.edu/research/index.html.

22. Noveck, "A Democracy of Groups," *First Monday* 10/11 (2005), http://www.firstmonday.dk/issues/issue10_11/noveck/index.html; Jefferson Center, *Citizens Jury® Handbook,* 2004, 3, http://www.jefferson-center.org/vertical/Sites/%7BC73573A1-16DF-4030-99A5-8FCCA2F0BFED%7D/uploads/%7B7D486ED8-96D8-4AB1-92D8-BFA69AB937D2%7D.pdf.

23. Bodies Electric LLC, "What is Unchat?," http://web.archive.org/web/20070101085058/http://www.unchat.com/unchat.html.

24. Don Adams and Arlene Goldbard, "Transforming Dialogue: Web Lab's Explorations at the Frontiers of Online Community," (report, MacArthur Foundation and Markle Foundation, 2000), http://www.weblab.org/sgd/evaluation.html.

25. Computer Professionals for Social Responsibility Public Sphere Project, "About e-Liberate: Support for Online Deliberation," 2004, http://trout.cpsr.org/program/sphere/e-liberate/about.php.

26. Information Renaissance, "Information Renaissance Model for Online Dialogues," http://www.info-ren.org/what/dialogues_background.shtml.

27. Carolyn J. Lukensmeyer and Steve Brigham, "Taking Democracy to Scale: Creating a Town Hall Meeting for the Twenty-First Century," *National Civic Review* 91, no. 4 (Winter 2002): 353–60, http://www.ncl.org/publications/ncr/91-4/ncr91-4_chapter6.pdf.

28. Jim Rough, "The Wisdom Council: A Strategy for We the People to Create True Democracy," Center for Wise Democracy, http://www.wisedemocracy.org/papers/Localdemocracy.html.

29. Robert D. Carlitz and Rosemary W. Gunn, "Online Rulemaking: A Step Toward E-Governance," *Government Information Quarterly* 19, no 4 (2002): 389–405.

30. Organization for Economic Co-operation and Development, "OECD Policy Brief: Engaging Citizens Online for Better Policy-Making," *OECD Observer* (policy brief, 2003), http://www.oecd.org/dataoecd/62/23/2501856.pdf.

31. Nicholas W. Jankowski and Renee Van Os, "Internet-Based Political Discourse: A Case Study of Electronic Democracy in Hoogeveen," in *Democracy Online: The Prospects for Political Renewal Through the Internet,* ed. Peter M. Shane, (New York: Routledge, 2004), 183–84.

32. Beth Simone Noveck, "Unchat: Democratic Solution for a Wired World," in *Democracy Online,* 32–33, http://www.nyls.edu/docs/noveck_unchat.pdf.

33. Coleman and Gøtze, "Bowling Together," 18.

34. Don Adams and Arlene Goldbard, "Transforming Dialogue: Web Lab's Explorations at the Frontiers of Online Community. An Evaluation Report for Web Lab," (report, MacArthur Foundation and Markle Foundation, 2000) http://www.weblab.org/sgd/ALS_Final_10500.pdf, 11–13; Denise Caruso, "Improving Dialogue on the Internet," *The New York Times,* July 5, 1999, http://www.nytimes.com/library/tech/99/07/biztech/articles/05digi.html.

35. Bodies Electric LLC, "What is Unchat?"

36. Computer Professionals for Social Responsibility Public Sphere Project, "About e-Liberate."

37. Clay Shirky notes, "The communities that thrive [online] violate most or all of the earlier assumptions. Instead of unlimited growth, membership, and freedom, many of the communities that have done well have bounded size or strong limits to growth, non-trivial barriers to joining or becoming a member in good standing, and enforceable community norms that constrain individual freedoms. Forums that lack any mechanism for ejecting or controlling hostile users . . . have often broken down under the weight of users hostile to the conversation." Clay Shirky, "Social Software and the Politics of Groups," Clay Shirky's Writings About the Internet, March 9, 2003, http://shirky.com/writings/group_politics.html.

38. Coleman and Gøtze, "Bowling Together," 13.

39. Thomas C. Beierle, "Digital Deliberation: Engaging the Public Through Online Public Dialogues," in Shane, *Democracy Online,* 157–58.

40. Lukensmeyer and Brigham, "Taking Democracy to Scale," 352; Organization for Economic Co-operation and Development, "Engaging Citizens Online."

41. Coleman and Gøtze, "Bowling Together," 13.

42. Barber, *Strong Democracy: Participatory Politics for a New Age* (Berkeley: University of California Press, 1984), 261–312.

43. Touraine, *Critique of Modernity* (Oxford, UK: Blackwell, 1995).

44. Habermas, *The Theory of Communicative Action (Vol. 2): Lifeworld and System* (Boston: Beacon Press, 1987), 392.

45. Manuel Castells, *The Power of Identity* (Oxford, UK: Blackwell Publishers, 1997); Nick Dyer-Witheford, *Cyber-Marx: Cycles and Circuits of Struggle in High-Technology Capitalism* (Chicago: University of Illinois Press, 1999); Alberto Melucci, *Challenging Codes: Collective Action in the Information Age* (Cambridge: Cambridge University Press, 1996).

46. Lauren Langman, Douglas Morris, and Jackie Zalewski, "Cyberactivism and Alternative Globalization Movements," in *Emerging Issues in the 21st Century World-System,* ed. Wilma A. Dunaway, 218–235 (Westport, CN: Greenwood Press, 2003).

47. On the Zapatista movement, see Harry Cleaver, "The Chiapas Uprising and the Future of Class Struggle in the New World Order," *Riff-Raff: attraverso la produzione sociale* (Padova, Italy: 1994), 133–145, http://libcom.org/library/chiapas-uprising-future-class-struggle-new-world-order-cleaver; John Arquilla and David Ronfeldt, *The Advent of Netwar* (monograph report: RAND, 1996); David Ronfeldt, Jon Arquilla, Graham Fuller, and Melissa Fuller, *The Zapatista "Social Netwar" in Mexico* (monograph report, RAND, 1998). On Indonesia, see Merlyna Lim, *@rchipelago Online: The Internet and Political Activism in Indonesia,* (PhD dissertation, University of Enschede, 2005).

48. MoveOn.org is perhaps the most famous online social movement organization in history. It made its debut in 1998 by launching an online petition against the impeachment of Clinton. It became famously known worldwide in 2001 with its online peace campaign following the WTC attack on September 11, 2001, which was quickly signed by more than half a million people. For further information about MoveOn.org, see http://www.moveon.org/about.html and http://en.wikipedia.org/wiki/MoveOn.org.

49. Cleaver, "The Chiapas Uprising."

50. Harry M. Cleaver, "Computer-linked Social Movements and the Global Threat to Capitalism," http://www.eco.utexas.edu/~hmcleave/polnet.html; Adrienne Russell, "Myth and the Zapatista Movement: Exploring a Network Identity," *New Media and Society* 7, no. 4 (2005), 572.

51. Viola Krebs, "The Impact of the Internet on Myanmar," *First Monday* 6, no. 5 (2001), http://firstmonday.org/issues/issue6_5/krebs/index.html.

52. Stefano Baldi, "The Internet for International Political and Social Protest: the case of Seattle," Research Paper No. 3 (Rome: Policy Planning Unit of the Ministry of Foreign Affairs of Italy, 2000).

53. Lim, *@rchipelago Online,* 183.

54. Rajiv Chandrasekaran, "Philippine Activism, At Push of a Button: Technology Used to Spur Political Change," *Washington Post,* December 10, 2000, A44; Vicente Rafael, "The Cell Phone and the Crowd: Messianic Politics in the Contemporary Philippines," *Public Culture* 15, no 3 (2003).

55. Rob Stuart and Jed Miller, "The Net Works: Prospects for Advocacy and Mobilization Online," (report, The E-Volve Foundation, 2003), http://www.jedmiller.com/graphics/netadvocacy_v2_04.pdf.

56. Meikle, *Future Active: Media Activism and the Internet* (London: Routledge, 2002); Marshall McLuhan, *The Medium is the Message* (New York, Bantam Books, 1967), 74–75.

57. Wikipedia contributors, "Blog," Wikipedia, The Free Encyclopedia, http://en.wikipedia.org/wiki/Blog.

58. A 2003 Pew Internet survey reports that only 4 percent of Americans reported going to blogs for information and opinion. Lee Rainie, Susannah Fox, and Deborah Fallows, "The Internet and the Iraq War: How Online Americans Have Used the Internet to Learn War News, Understand Events, and Promote Their Views," (report, Washington, DC: Pew Internet & American Life Project, 2003), 5.

59. Amanda Lenhart and Susannah Fox, "Bloggers: A Picture of the Internet's New Storytellers," (report, Washington, DC: Pew Internet and American Life Project, 2006), i, http://www.pewinternet.org/pdfs/PIP%20Bloggers%20Report%20July%2019%202006.pdf.

60. A rough ranking of Boing Boing's popularity can be found at the Technorati list of most popular blogs http://technorati.com/pop/. Note also that the Huffington Post, a political blog, is among the top five.

61. Sitemeter, *Daily Kos: Site Summary,* retrieved on January 11, 2007, http://www.sitemeter.com/?a=stats&s=sm8dailykos.

62. Mike Webb, the *Nation* Publicity Directory, Press Release, "Victor Navasky to Step Down from Nation Helm After 28 Years," November 7, 2005, http://www.commondreams.org/news2005/1107-04.htm.

63. Dan Gillmor, *We the Media* (Cambridge, MA: O'Reilly, 2004).

64. Matthew Hindmann, Kostas Tsiotsiouliklis, and Judy Johnson, "Googlearchy: How a Few Heavily linked Sites Dominate Politics Online," paper presented at the annual meeting of the American Political Science Association, Philadelphia, PA, August 2003.

65. Shirky, "Power Laws, Weblogs, and Inequality."

66. Adamic and Glance, *The Political Blogosphere and the 2004 U.S. Election: Divided They Blog,* March 4, 2004, http://www.blogpulse.com/papers/2005/AdamicGlanceBlogWWW.pdf.

67. Sunnstein, *Republic.com;* Anthony G. Wilhelm, *Democracy in the Digital Age: Challenges to Political Life in Cyberspace* (London, UK: Routledge, 2000).

68. Techmeme Leaderboard, http://www.techmeme.com/lb.

69. Gibson, "God's Little Toys: Confessions of a Cut & Paste Artist," *Wired* 13.07 (2005), http://www.wired.com/wired/archive/13.07/gibson.html.

70. Dybwad, “Approaching a definition of Web 2.0,” The Social Software Weblog, posted September 29, 2005, http://socialsoftware.weblogsinc.com/2005/09/29/approaching-a-definition-of-web-2-0/.

71. Chris Morris, *Bushwacked2,* http://www.warprecords.com/news/?offset=0&ti_id=573.

72. Jonathan McIntosh, *Hummertruth,* http://www.capedmaskedandarmed.com/video/hummertruth.mov.

73. Iman Foroutan, *Zendanie Siasi,* http://www.democracyforiran.de/zendani256k.wmv; Alex Chan, *French Democracy* (French, 2005), http://www.machinima.com/films.php?download=1407.

74. Jenkins, *Convergence Culture* (New York: New York University Press, 2006).

75. Ivan Illich, *Tools for Conviviality,* 78–79.

4

Infrastructure: Network Neutrality and Network Futures

François Bar, Walter Baer, Shahram Ghandeharizadeh, and Fernando Ordonez

More than just a new medium, the Internet is fast becoming our primary communication infrastructure, progressively supplanting older radio, telephone, and cable television networks. In the Net's early days, during the 1970s and 1980s, it mainly provided support for text exchanges through e-mail, broadcast listservs, and text-based precursors to the Web such as Gopher servers. The widespread traffic of images came next, with easier media attachments to e-mail and the introduction of Mosaic, the first popular graphical Web browser, in 1993. Today, with broadband access widely available, the Internet is commonly used to transmit audio, including a growing share of our phone conversations through VoIP, streamed radio programs, podcasts, and recorded music. Video, too, is being broadcast over the Net, from traditional television programs and films to video blogs, home movies, and creations made by a multitude of emerging amateur producers. Thus, a single communication infrastructure has progressively absorbed a multitude of media streams that once each required specialized networks.

This is not to say that old networks are simply discarded. Throughout this book, we have observed how transformations in place, culture, and politics build on existing historical conditions. The case of infrastructure is no different. Existing telephone copper wires, cable television's coaxial lines, long-haul optical fibers, and satellite and microwave-radio links are being folded into the Internet as telecommunications companies and cable carriers convert their respective networks to Internet Protocol (IP)—the set of data transmission conventions that allow communications across the various parts of the Internet. In fact, rather than a separate physical infrastructure, the Internet is primarily a virtual network—the assemblage of a multitude of transmission and routing facilities tied together by the IP's common software "glue." As a result, the Internet is perhaps best understood in its original, unabbreviated sense, as an *Internetwork,* or an agglomeration of separate networks that agree to connect to each other and exchange traffic through gateways where they speak the IP lingua franca. As an Internetwork, the Internet differs from traditional telephone and cable networks in two fundamental respects—its decentralized governance and its E2E architecture—and yet, relying on telephone and cable networks for the last mile of connectivity, the Internet is subject to economic and political pressure from established communications companies.

Decentralized governance means that no single organization is in charge of managing the Internet, which is in stark contrast with traditional telephone or television networks. In particular, individual networks can become part of the Internet as soon as they find an existing member of the Net that agrees to connect with them through a gateway to exchange traffic. Connected to one node, they are connected to the entire Net. Historically, this allowed the Internet's spectacular growth, as more and more network operators chose to join.

A brief glance at the history of the Internet illuminates the exponential nature of growth that this decentralized system allowed. Initially, the IP architecture was worked out in the ARPANET (Advanced Research Projects Agency Network), a military-sponsored experiment, during the early 1970s. It was notably expanded in the late 1970s and 1980s to support computer-intensive research through the government-sponsored NSFNet (National Science Foundation Network), an IP-based network linking universities. By the end of the decade, a number of corporations began to use the same networking approach to build their internal corporate networks, and a multitude of private Internet Service Providers (ISPs) started to offer dial-up Internet access over telephone companies' lines. During the 1990s the Internet became a mass medium, propelled by easy-to-use, multimedia content available through the World Wide Web and the absorption of consumers from earlier, largely self-contained networks such as CompuServe, Prodigy, and America Online (AOL). Throughout, the Internet's expansion did not require the blessing of centrally-controlled telephone or television networks, but instead proceeded in a decentralized fashion as increasing numbers of private and public operators adopted the new networking model and peered with existing participants to join the Internetwork.[1]

Another unique feature of the Internet is its E2E architecture, which further fueled this success because it enabled the deployment of a communication infrastructure that did not predetermine how it would be used, thus opening the Internet to wide-ranging experimentation and innovation. The E2E model calls for processing information in the devices connected to the ends of the network whenever possible, while the Internet itself remains a dumb network,

chapter explores these trade-offs and the impact they could have on the evolution of our communication infrastructure and the activities it will support.

Speed Bumps on the Road Toward Ubiquitous Broadband

Some of the tensions underlying the transition toward the next generation Internet infrastructure are universal. Network owners in all countries, whether telecommunications carriers, cellular providers, or cable systems, are moving toward the deployment of integrated broadband IP infrastructures for all communication services (voice, data, and video) to the home. In doing so, they need to secure funds to finance the upgrades to infrastructure and must set up sustainable business models for operation in light of falling costs for traditional cash generators such as long-distance tolls.

Nevertheless, the American context is unique and explains the specifics of the domestic policy debate. Three features of this context stand out: the eroding U.S. position in broadband worldwide, the enduring structure of the U.S. local access market as a duopoly, and the lack of a national broadband policy.

The local access network, what the phone companies used to call the last-mile connection between long-distance networks and customers premises, be they residences or corporate campuses, is the critical bottleneck for broadband. Indeed, today there is abundant bandwidth available in the Internet's backbone, in part as a result of exuberant investment in cross-country optical-fiber links during the dot-com boom. Some of these optical fibers are still dark, awaiting the installation of lasers to light them up so they can carry information, but the potential is there. Similarly within customer premises, business and residential alike, high-speed networks are commonplace, either in the form of Ethernet's ubiquitous blue Cat 5 cables or wireless Wi-Fi networks. The missing link for true broadband deployment is the connection between the two networks.

The United States led the development of Internet technology and its early deployment into a widely used infrastructure. In the current transition to universal broadband, however, other countries have taken the lead. OECD data (figure 4.1) shows the United States ranking fifteenth among OECD countries in broadband subscribers per one hundred people as of December 2006.[5]

Analysts attribute the broadband lag in the United States to a combination of demographic, economic, institutional, and policy factors. Broadband penetration is positively correlated with population density. The United States has lower population density than most other OECD countries, which increases the average length and cost of American broadband access lines. Dial-up Internet

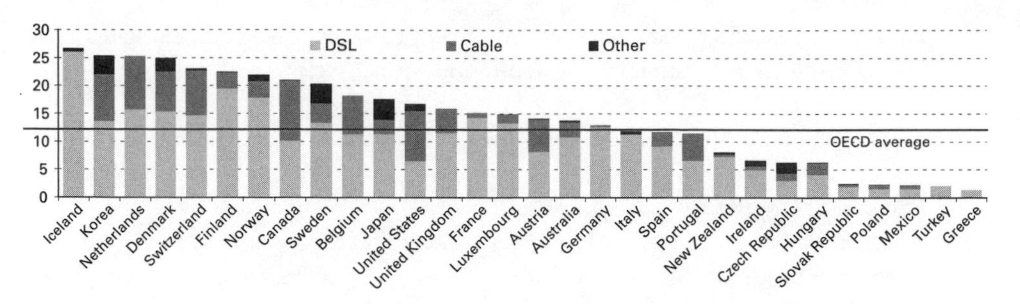

Figure 4.1 OECD broadband subscribers per 100 inhabitants, December 2005. http://www.oecd.org/document/39/0,2340,en_2649_201185_36459431_1_1_1_1,00.html

access may also be a better substitute for broadband in this country, where local calls are generally free, than in other countries where local calls are metered.[6]

Moreover, unlike Japan, Korea, and some European countries, the United States lacks a national policy to spur broadband uptake. Compounding this, the United States also has a telco/cable duopoly with relatively weak broadband competition from other providers.[7] Studies within and among OECD countries generally conclude that more competition results in lower prices and greater broadband penetration.[8]

Government policy to encourage and subsidize investment in fiber-optic networks has spurred big broadband—above 20 megabits per second (Mbps)—development in Japan and South Korea. In 1995, the South Korean government sponsored construction of a nationwide, high-capacity-fiber broadband network that could be used by any telecom carrier. The Korean government also subsidized loans to broadband providers as well as user purchases of personal computers. Moreover, government unbundling policies that require incumbents to allow competitors to use their local access lines have enabled big broadband competition from companies such as Yahoo! BB in Japan and Hanaro Telecom in South Korea.[9] This international situation provides the backdrop for the current U.S. policy discussion about the future of the Net. All sides agree that access to an advanced communication infrastructure is essential to the future social and economic well-being of the country. There is substantial disagreement, however, about how best to achieve this goal.

In the United States, each local market is at best served by a duopoly: customers who want broadband Internet access can choose between telco-provided DSL (Digital Subscriber Line) service or cable-modem service from their local cable operator. In large portions of the country, particularly in rural areas, customers only have one broadband option, or even none. Where other

Table 4.1 Broadband price and speed by selected OECD country and provider, September 2005

Country and Providers	Type	Monthly price ($)	Max data speed (Mbps) Down	Up	$ per month per Mbps
Canada					
Incumbent telco (Bell Canada)	DSL	49	5	0.8	9.8
Cable provider (Cogeco)	Cable	68	10	1	6.8
Alternate provider (Aliant)	DSL	44	5	0.6	8.8
France					
Incumbent telco (France Telecom)	DSL	49	18	1	2.7
Cable provider (Noos)	Cable	43	10	NA	4.3
Alternate provider (Free Telecom)	DSL	37	20	1	1.8
Germany					
Incumbent telco (Deutsche Telecom)	DSL	43	6	0.6	7.1
Cable provider (Kabel Deutschland)	Cable	37	6	0.4	5.9
Alternate provider (Arcor)	DSL	37	6	NA	6.1
Japan					
Incumbent telco (NTT West)	Fiber	37	100	100	0.4
Cable provider (J:COM)	Cable	52	30	2	1.7
Alternate provider (Yahoo! BB)	Fiber	39	100	100	0.4
Korea					
Incumbent telco (KT)	Fiber	38	100	100	0.4
Cable provider (C&M)	Cable	29	5	NA	5.8
Alternate provider (Hanaro)	Fiber	46	50	50	0.9
United States					
Incumbent telco (AT&T)	DSL	39	3	0.4	13.1
Incumbent telco (Verizon)	Fiber	45	15	2	3.0
Cable provider (Comcast)	Cable	72	6	0.8	12.0
Alternate provider (Mstar)	Fiber	42	15	15	2.8

Source: OECD, Multiple Play: Pricing and Policy Trends, 2006; Verizon, 2006.

Note: Table 4.1 compares broadband speed and prices as of September 2005 for incumbent telco, cable, and alternative Internet service providers (ISPs) in the United States, Canada, France, Germany, Japan, and Korea (OECD, 2006). The table clearly shows the difference between 100 Mbps big broadband over fiber in Japan and Korea, and mini broadband over DSL (digital subscriber line) and cable in Europe and North America. In the United States, Verizon is aggressively promoting its FiOS (Fiber Optic Service) broadband service over fiber with speeds up to 30 Mbps, but at prices significantly higher per Mbps than what is offered in Korea and Japan. AT&T has recently introduced its fiber-based U-Verse television and data service in San Antonio, Texas, but its data speeds and prices basically match what is already available from DSL and cable in that market.

alternatives exist, they come mostly from companies that provide Internet service over telephone lines they lease from the incumbent telco. For all practical purposes, the two members of this TCNO duopoly control the current availability and future evolution of the local distribution network, key to the connection between high-capacity Internet backbone and customer premises. So long as that situation persists, they jointly hold the key to the future of the country's broadband Internet infrastructure.

For big broadband to become available throughout the United States, and for the country to reverse its downward slide in the broadband rankings, there are two fundamental options. The first is for the TCNOs to make significant investments to upgrade their networks; the second is for an alternative broadband infrastructure to be deployed that can challenge the existing duopoly.

The TCNOs claim that making the investments needed to offer ubiquitous big broadband as opposed to current DSL and cable mini-broadband (5 Mbps or less downstream; much less upstream) would only make financial sense if they were allowed greater control over the applications and content delivered over these networks. In particular, the telcos point to obstacles inherited from their history as common carriers that require them to transmit all messages and data alike (e-mail data bits just like high-definition video frames, traffic from partners just like traffic from competitors). According to them, unleashing adequate investment for the broadband infrastructure requires a break from the Internet's tradition of decentralized control and E2E architecture. (See box 4.1.)

The second option, the deployment of an alternative broadband infrastructure, has been the elusive goal of U.S. telecommunication policy over the past two decades. This was, in particular, the underlying goal of the Telecommunications Act of 1996, which structured a set of incentives for the telecommunications companies to gain greater regulatory freedom as they allowed competitors to enter their local markets. These efforts have largely failed, and to date no infrastructure builder has emerged that can credibly compete with the TCNOs. Over the past few years, however, a new candidate has come forward with the advent of wireless data networks. Promoted in particular by the newfound availability of unlicensed radio spectrum, technologies such as Wi-Fi and WiMAX suggest that new infrastructure models could provide the basis for an entirely new last-mile broadband infrastructure. This could come in the form of wireless networks run by for-profit firms or local government operators. More dramatically, some also foresee the advent of ad hoc networks, where wireless-enabled devices connect to one another when they come within radio range, creating a decentralized mesh without operators. We review these options in more detail later on.

Box 4.1a TCNO Views on Net Neutrality

From BellSouth, "Overview of Net Neutrality," (2006), http://www.democraticmedia.org/PDFs/BellSouthNetNeutral.pdf.

Under BellSouth's view of net neutrality, the essential consumer protection is clear disclosure in the service plan agreement. In addition, providers must have substantial flexibility in managing their broadband networks and in structuring business arrangements with customers and content providers, and this flexibility must be recognized and preserved in any mandated net neutrality arrangements.

First, broadband network providers should be able to take steps to ensure network security and preserve the integrity of their overall systems. For example, providers should be able to implement measures to prevent spamming, the release of viruses on their networks, and the use of their networks for unlawful purposes.

Second, broadband network providers should be able to manage bandwidth. Although it would be irrational for any broadband network provider to aggressively limit bandwidth consumption, since they make money by ensuring greater amounts of network utilization, providers should be able to curb network usage (such as peer-to-peer file sharing) that consumes a disproportionate amount of bandwidth and may adversely impact other network users. Bandwidth management is a particularly important issue in the context of IPTV [Internet Protocol Television].

Third, broadband network providers should be able to offer different plans that feature enhanced levels of service or that promote their own brand names and products or the services of selected vendors. For example, BellSouth should be able to enter into arrangements with content providers by which the content provider pays for special treatment, such as preferential listing or faster downloads from that provider's website or receiving higher quality of service.

Box 4.1b TCNO Views on Net Neutrality

From Edward Whitacre, CEO, SBC (now AT&T), "At SBC, It's All About 'Scale and Scope'," Business Week Online, November 7, 2005, http://www.businessweek.com/@@n34h*IUQu7KtOwgA/magazine/content/05_45/b3958092.htm.

How do you think [content providers are] going to get to customers? Through a broadband pipe. Cable companies have them. We have them. Now what they would like to do is use my pipes free, but I ain't going to let them do that because we have spent this capital and we have to have a return on it. So there's going to have to be some mechanism for these people who use these pipes to pay for the portion they're using. Why should they be allowed to use my pipes?

The Internet can't be free in that sense, because we and the cable companies have made an investment and for a Google or Yahoo! or Vonage or anybody to expect to use these pipes [for] free is nuts!

The Debate Over Network Neutrality

There is considerable debate about whether the Internet's upgrade to ubiquitous big broadband can proceed within the traditional decentralized, E2E framework. In particular, the TCNOs who will have to invest in this infrastructure upgrade claim they can only justify this next round of investment if they are able to capture a portion of the revenues associated with the new broadband applications—in particular video distribution. In order to do this, they seek to shape traffic over their network so they can deliver better performance to those (content/application suppliers or consumers) willing to pay for it. TCNOs argue that unless they can manage traffic and applications in such a manner that they can reap economic benefits, they will have no incentives to invest in network upgrades and, thus, have little reason to expedite the move toward ubiquitous and affordable broadband for all end users. Of course, network owners already manage traffic to some extent, for example, when they allocate available bandwidth among neighbors. But the kind of management they yearn for would allow them to differentiate traffic on the basis of business relationships with content providers and consumers. For example, they might make YouTube videos flow more smoothly if YouTube gave them a cut of the related advertising revenues, or provide lower latency (faster reaction time) to World of Warcraft players in exchange for a share of the game subscription fees.

In response, a coalition of content and application providers (including Amazon.com, eBay, Google, Intel, Microsoft, and Yahoo!) and consumer protection organizations have argued that by unduly favoring TCNO-owned content, this would be the end of the Internet as we know it. They have proposed network neutrality principles to guarantee that network owners do not treat different traffic flows differently, whether on the basis of fees paid by those exchanging traffic or according to what application they use. They see network neutrality as a way to ensure overall economic and social benefits in light of increased TCNO market power (see box 4.2). Network neutrality advocates argue two fundamental points. First, information networks should preserve the E2E Internet architecture and be as neutral as possible among competing content, applications, and services. Second, if and when it is necessary, government should intervene to promote or preserve the neutrality of these networks.[10] The FCC did, in fact, adopt a broadband policy statement in August 2005 that consumers are entitled to their choice of lawful Internet content, applications, services, and devices (see box 4.3); but the FCC has not adopted rules to enforce these principles.

This underlying tension about how much control infrastructure owners should be allowed to wield over the communication activities carried by their infrastructure is not new.[11] As with Carterfone or the FCC's Computer

Box 4.2 Google's Support for Network Neutrality

From "Prepared Statement of Vinton G. Cerf, Vice President and Chief Internet Evangelist, Google Inc.: U.S. Senate Committee on Commerce, Science, and Transportation, Hearing on 'Network Neutrality,' February 7, 2006," http://commerce.senate.gov/pdf/cerf-020706.pdf.

The Internet's open, neutral architecture has proven to be an enormous engine for market innovation, economic growth, social discourse, and the free flow of ideas. The remarkable success of the Internet can be traced to a few simple network principles—end-to-end design, layered architecture, and open standards—which together give consumers choice and control over their online activities. This "neutral" network has supported an explosion of innovation at the edges of the network, and the growth of companies like Google, Yahoo!, eBay, Amazon, and many others. Because the network is neutral, the creators of new Internet content and services need not seek permission from carriers or pay special fees to be seen online. As a result, we have seen an array of unpredictable new offerings—from Voice-over-IP to wireless home networks to blogging—that might never have evolved had central control of the network been required by design.

Allowing broadband carriers to control what people see and do online would fundamentally undermine the principles that have made the Internet such a success. For the foreseeable future most Americans will face little choice among broadband carriers. . . . Phone and cable operators together control 98 percent of the broadband market, and only about half of consumers actually have a choice between even two providers. Unfortunately, there appears to be little near-term prospect for meaningful competition from alternative platforms. As a result, the incumbent broadband carriers are in position to dictate how consumers and producers can use the on-ramps to the Internet.

A number of justifications have been created to support carrier control over consumer choices online; none stand up to scrutiny. Open-ended carrier discrimination is not needed to protect users from viruses, stop spam, preserve network integrity, make VoIP or video service work properly—or even insure that carriers are compensated for their broadband investments. In particular, we firmly believe that carriers will be able to set market prices for Internet access and be well-paid for their investments—as broadband carriers in other countries have successfully done.

Even as we welcome the deregulation of our telecommunications system, we must preserve some limited elements of openness and non-discrimination that have long been part of our telecommunications law. In this regard, Google supports tailored, minimally-intrusive safeguards to promote net neutrality.

Box 4.3 FCC Policy Statement on Network Neutrality
From Federal Communications Commission, Policy Statement (FCC 05-151), adopted August 5, 2005.

In August 2005, The Federal Communications Commission adopted a policy statement "that outlines four principles to encourage broadband deployment and preserve and promote the open and interconnected nature of public Internet:

(1) consumers are entitled to access the lawful Internet content of their choice;

(2) consumers are entitled to run applications and services of their choice, subject to the needs of law enforcement;

(3) consumers are entitled to connect their choice of legal devices that do not harm the network; and

(4) consumers are entitled to competition among network providers, application and service providers, and content providers.

Although the Commission did not adopt rules in this regard, it will incorporate these principles into its ongoing policymaking activities. All of these principles are subject to reasonable network management."

Inquiries in previous incarnations of this debate, there are two categories of issues. The first is about gatekeeping; static, immediate, and highly visible, it captures much of the attention. The second is about future evolution; dynamic, longer term, and unknown, it is much harder to measure its impact.

The gatekeeping debate revolves around the fact that TCNOs want to charge a premium to those who want to transmit their content over reliable broadband streams (broadcasters, game servers, etc.). In order to deliver high-quality broadband streams to their highest-paying customers, TCNOs claim they need the ability to shape traffic—to decide which packets have priority, and which communication flows get reserved bandwidth. With the ability to shape traffic, they also get control over content flows, ranging from small tweaks in how quickly Web pages load (for example, making it faster for content providers willing to pay more) to outright censorship by blocking certain categories of content or certain applications (for example, redirecting Internet users toward online stores with whom they have a partnership or preventing them from using the VoIP services of their competitors).

Other competitive issues can arise if the TCNOs configure their network for a specific kind of communication activity and as a result hinder other kinds of traffic. For example, a network optimized for video broadcasting downstream, with very little upstream capacity, would inhibit VoIP services. Network operators could also privilege the traffic of their content partners and hinder other flows. In particular, this is most likely to affect amateur or small-scale content and application providers, who are less likely able to set up partnerships with large TCNOs. One strong countervailing force is that carriers want to deliver what they call the "triple play," to hold on to their customers by delivering all their video, voice, and data traffic.[12] Decisions in this regard are likely to come down to their assessment of whether they stand to gain more from carrying more traffic or from prioritizing certain traffic.

Whatever the case, such gatekeeping threats are easily detectable: one can imagine consumer advocates benchmarking performance on different sites. At the extreme, outright blocking of certain content or applications is unlikely to go unnoticed. In a rare case where its service was blocked by Madison River Communication, VoIP provider Vonage complained and the FCC fined the telco and forced it to stop the practice.[13] So although there is a threat at the gatekeeping level, there are also existing checks against it.

The second kind of danger will emerge as the network evolves and, we believe, is more insidious and potentially more consequential. When infrastructure owners design their network to favor specific applications and prioritize certain kinds of traffic, they inevitably limit—or outright prevent—experimentation with alternatives. For example, if the network owners decide to optimize their infrastructure to best deliver asymmetric video flows, the resulting network will inevitably be less suited to experimentation with new kinds of symmetric applications. One could argue that the innovation process leading to the emergence of a participatory, collaborative, media-rich communication platform, what many observers refer to as Web 2.0, is greatly encouraged by the availability of a neutral, E2E Internet infrastructure.[14] If the TCNOs decide to optimize the next generation infrastructure for video distribution, what are the paths of innovation that will remain open for Web 3.0? Proponents of network neutrality argue that to continue to have the level of innovation that has made the Internet a success, the network infrastructure must preserve its E2E architecture.

By contrast with gatekeeping, the consequences of evolution are much harder to monitor and assess. With respect to future evolution, the opportunity cost of not requiring network neutrality lies in avenues not explored. It is therefore impossible to know what we might miss. Network neutrality opponents would also argue that different kinds of innovation will be encouraged by allowing infrastructure owners to shape their networks in favor of certain categories of applications. For example, an Internet optimized for broadcast would probably lead to greater innovation in technologies and services supporting asymmetric, one-to-many communication patterns, perhaps even forcing a return to the kind of mass-media culture that existed prior to the spread of Internet. In the end, this may come to a trade-off between favoring innovation along a preselected path (or a few paths) versus encouraging exploration of a broad variety of evolutionary paths.

Overall, whether we consider the static impact of gatekeeping or the dynamic implications of favoring alternative evolutionary paths, there may not be a clear-cut best choice for the future of our communication infrastructure. Rather, there are a number of trade-offs to be considered, and the decision will have to be political rather than technical. In order to explore some of these trade-offs, we developed three scenarios about possible futures for the U.S. network infrastructure. These scenarios, which follow this chapter and are available on the Web at http://networkedpublics.org/conference/infrastructure_videos, were prepared for the April 2006 Networked Publics Conference to stimulate discussion about possible network futures. They were not meant to describe the entire range of possibilities, but rather to help tease out some of the important dimensions of this debate about the network's future and explore ways to resolve the associated trade-offs. Our three scenarios differ along three dimensions in particular: the extent to which the infrastructure is specialized or generic; the locus and extent of control over patterns of communication and content; and the forms and extent of privacy and security. Each relates to basic principles about the future of the communication infrastructure, and the choices made will powerfully shape how networked publics emerge, as explored in the other chapters in this book.

The first scenario, *Neutral Net,* assumes passage of comprehensive network neutrality rules, forcing any infrastructure owner to let others use the network on a nondiscriminatory basis. In this future, carriers would likely adjust their business models to focus on bit carriage, multiple organizations would provide competing communication services over that open infrastructure, and end users would engage in wide-ranging experimentation with applications and content. The result would likely be an infrastructure able to support a large variety of communication patterns and applications, though perhaps less directly optimized to the needs of any single one of them. Infrastructure owners would not be able to favor certain content or applications over others. Privacy and security concerns would likely have to be addressed by legislation, along a model that could be an extension of traditional common carrier principles.

By contrast, the second scenario, *TCNOtopia,* assumes that lawmakers reject any form of network neutrality and give TCNOs free reign to control and shape traffic as they see fit. In that future, network owners would be able to optimize their network architecture to deliver high performance for the applications and content that generate revenues for them or their business partners, without having to provide comparable performance or access to their competitors' applications and content. They would be in charge of the shape and evolution of the communication infrastructure, including implementation of privacy and security features that enable better control of outside threats to users, such as spyware, spam, and viruses. On the other hand, network owners would have greater freedom to exploit their detailed knowledge about customers' communication patterns for commercial gain.

The third scenario, *AutoMata,* explores a future in which an alternative broadband network emerges separately from the existing TCNO infrastructures. This new wireless mesh network would develop from the spontaneous agglomeration of devices, primarily in cars and other vehicles that are able to communicate as soon as they find themselves within radio range of each other. There would be no communication infrastructure per se, but rather organic, ad-hoc networks of radio devices that create multi-hop pathways for exchanging information among users. Control would be widely distributed in this decentralized network, and experimentation with new forms of content and applications is broadly empowered. However, this might result in a network highly vulnerable to privacy and security threats, a wireless Wild West of sorts, where individuals must shoulder the responsibility to protect themselves from harm and abuse.

In the next section, we examine some of the technological factors that influence these possible futures, as well as alternative possibilities for the emergence of a third broadband infrastructure that can effectively compete with the existing telephone and cable networks.

Technical Factors and Trade-offs Driving Broadband Evolution

Discussions of broadband deployment tend to focus on bandwidth (used here as synonymous with data speed) alone. Over time, users and application/service providers will demand faster broadband connections, especially for music and video applications. With audio and video clips, higher bandwidths are essential to stream data in support of a hiccup-free display. High-definition video in particular will require big broadband speeds greater than 20 Mbps.

However, other technical characteristics of the future communication infrastructure can be equally important, especially as the Internet moves beyond the current dominance of Web-based applications. In addition to bandwidth, there are four other key dimensions that can be particularly critical to the communication practices (such as multiplayer games, pervasive media, or user-produced content) that are highlighted in the other chapters of this book: low latency, symmetry, ubiquity and affordability, and mobility.

Latency on the Internet or in other telecommunications networks is the amount of time, generally measured in milliseconds, it takes to get a response to a request.[15] Latency depends not only on the bandwidth, or speed, of the communications lines between sender and receiver, but also on the transit time over the network and the computer processing time within the network and at both ends. For example, latency between two points on a satellite link will be one to two seconds longer than on a terrestrial link, even if the link bandwidths are the same. And latencies may differ among similar packets that take different routes on the Internet, especially if some packets go through lower speed lines or significantly more routers than others.

Low latency is critical for synchronous communication (e.g., live voice or videoconferences and live video broadcasts) and for real-time collaboration (e.g., multiplayer games or concurrent engineering). This is why satellite Internet is a poor choice for multiplayer games or videoconferences. It is also one reason why telco and cable operators want to manage Internet traffic actively within their networks, so they can prioritize delivery of packets that require low latency (and perhaps other packets for which they receive higher reimbursement). An E2E alternative is to accelerate deployment of big broadband at speeds greater than 20 Mbps in which latency becomes less of a problem.

For asynchronous communication, latency is much less critical and can be mitigated using intelligent traffic management techniques such as pre-staging of content.[16] This can be done either within the core network using carrier-controlled equipment or with user-owned servers and other equipment at the network "edges" that are not under direct carrier control.

Where latency describes the responsiveness of one's connection, symmetry refers to the ratio of bandwidth for uploading as compared to bandwidth for downloading. Thus far, DSL and cable broadband (and so far, incumbent telco broadband over fiber in the United States) have been designed to be asymmetric, with speeds typically eight to ten times faster for downloading than uploading (see table 4.1). While such asymmetry works well for Web browsing and similar applications where users receive many more bits than they generate, applications that involve real-time video streaming from multiple sources (such as videoconferencing as well as online-based applications) can require more symmetric download and upload capabilities.

The growth of amateur, DIY video content could also affect the need for symmetric distribution bandwidth. Most DIY production is supported well by today's asymmetrical networks. Amateurs upload their content at relatively low speed to well-connected servers from which others can download at higher speed. That is the concept of the "Google grid" and similar approaches that allow access to a grid within which users can store and publish. On the other hand, using P2P distribution (e.g., BitTorrent or P2PTV) to share content files demands more symmetric network capabilities.

Moreover, the Internet is not simply a production/publishing platform, but also a communication and collaboration channel (person-to-person, person-to-computer, computer-to-computer, thing-to-thing).[17] It is difficult today to estimate whether this traffic will be mostly symmetrical or asymmetrical in the future.

What of ubiquity? The examples of Japan and South Korea indicate that making big broadband widely available at affordable prices over fiber-optic access networks is technically quite feasible. But such connectivity throughout the United States would require new investment that would likely approach $100 billion. While the TCNOs continue to invest in extending their DSL and cable broadband networks, by 2010 fewer than 15 percent of U.S. households will have big broadband with the capabilities that are already available in Asia.[18]

Such new investment could be funded by telco and cable network operators (as in our TCNOtopia scenario), by local governments, by application/service providers (e.g., Google's partnership with municipal Wi-Fi deployment in the TCNOtopia scenario), or by end users (e.g., in a wireless mesh scenario such as AutoMata). However, the TCNOs say they would have little incentive to invest if government enforces network neutrality. Some on Wall Street also argue that end-user affordability will depend on network operators having multiple income streams, such as charging content providers for fast, low-latency distribution.

The arguments and trade-offs between accelerating investment in broadband networks and sustaining user innovation in a nondiscriminatory E2E environment are at the core of the current network neutrality debate.

Nor is it enough to offer ubiquitous Internet in homes and offices. As lifestyles and habits change, Internet users are demanding broadband connectivity in different locations or while they are moving from place to place. Many users now carry a portable device—such as a laptop, smart phone, or PDA—for movable access. With this device they connect to the Internet from school and work, different rooms at home, hotels, Internet cafés, airports, and other fixed locations. The broadband access technology (DSL, cable, fiber, or wireless) will differ from place to place, although most users expect performance comparable to what they get at their primary access location. Users currently expect less from mobile Internet access while on the go, since it is available today primarily from mobile phone networks with lower performance and higher cost than other broadband options. However, expectations for mobile broadband will undoubtedly increase as new devices and networks become available.

There are two technical ways to pursue each of these issues: brute force (e.g., by deploying massive bandwidth) or clever network engineering (e.g., virtual circuits (IPv6) and quality of service (QoS) features). The first is compatible with network neutrality, but the second is generally not, as it involves making particular, low-level modifications to the way the network operates.[19] In fact, the choice also demands decisions about whether to address these technical features within the core network or outside it with user-controlled devices at its edges. This involves important trade-offs such as: Who pays for the incurred costs? Who controls access to these features? Who can capture the revenues and the benefits resulting from their use?

In the short term, these trade-offs reflect bounded choices about how best to provide infrastructure for applications we know and understand. The longer-term dynamics associated with these trade-offs are much more complex because then the choices are among different innovation trajectories and different predictions about what kinds of innovation we want to encourage for future applications we cannot yet imagine nor understand.

To a large extent, this is a replay of debates from the 1970s and 1980s about telco plans for an intelligent network versus the E2E principles that guided the designers of the early Internet.[20] Of course, there is no way to tell what innovative opportunities were lost from not pursuing the intelligent network route. But as the wealth of media on the Internet today proves, there is plenty of evidence that E2E has yielded spectacular results.

Is There a Viable Third Infrastructure?

The current discussion about the future of broadband infrastructure plays out against the background of the current last-mile duopoly of TCNOs. Both telephone and cable network operators have largely similar visions of where they are headed and, as a result, the fundamental policy debate is whether they should be left alone, or whether government should impose some rules on how they build and operate their networks.

This debate would be dramatically different if there were a third viable broadband infrastructure platform beyond the control of the incumbent TCNOs. Many of the concerns about loss of user innovation and other disadvantages to users in a TCNO-dominated future would disappear if greater competition existed for broadband Internet access. Prospective candidates for competitive access include several approaches to broadband wireless, municipal or other government-owned fiber-optic networks, and user-owned access links. From a technical and economic perspective, we are less sanguine about the prospects for broadband access over electric power lines or satellite, although satellite access can be important in rural areas where other alternatives are not available.

Broadband wireless access

Wireless Internet access is available in several forms, including mobile phones, local Wi-Fi networks, direct point-to-point links using fixed-wireless technologies such as WiMAX, and satellite. Each has some advantages, but each faces significant problems in scaling up to a widely available and affordable broadband infrastructure. Finding a successful evolutionary path from today's limited range and data capacities to broadly interconnected networks of wireless devices is, from our viewpoint, the key to wireless becoming a third national broadband infrastructure.

Mobile phone networks serve more than two hundred million customers in the United States and are aggressively working to expand beyond narrowband voice and text services to deliver music, images, video, and high-speed data. Building on the huge infrastructure they already have in place, U.S. cellular carriers have selectively introduced 3G services that offer Internet download speeds of 300–700 Kbps (kilobits per second). Cellular 3G offers mobile Internet access at near-broadband speeds where coverage is available, but that coverage is limited and generally costs more than DSL and cable alternatives. Not surprisingly, demand up to now appears limited primarily to businesses and high-income consumers.

Moreover, from the perspective of this chapter, current 3G and other cellular services are not compatible with network neutrality. Cellular firms operate proprietary, closed systems with active network management and full control over the services offered. For example, mobile phones, PDAs, or other devices used for Internet access on one cellular network generally cannot be used on other networks. Cellular carriers, in fact, often disable device features that would enable users to receive music, videos, or interactive services from other than their own affiliated sources. It seems unlikely that these proprietary business models will shift significantly in the next several years, particularly since Verizon, AT&T, and other incumbent wire-line carriers dominate the U.S. cellular industry. Consequently, it is unclear when, or whether, cellular networks will become more than niche competitors for broadband Internet access.

Wi-Fi networks, based on the IEEE 802.11 technical standards developed in the 1990s and using unlicensed spectrum, have grown phenomenally over the past decade, both within homes and as hot spots in commercial and public spaces. Wi-Fi hubs currently have a limited range of, at most, a few hundred feet and offer symmetric data speeds on the order of tens of megabits, which are shared by all devices on the network. Technically, however, local Wi-Fi networks do not scale very well,[21] and they are subject to interference problems.[22] Moreover, as presently implemented, they generally rely on DSL or cable connections to the Internet. Wi-Fi usage thus is covered by DSL or cable terms of service, which generally forbid unrestricted sharing as well as limited upstream data rate provisions.

Unlike 3G, however, it is possible to imagine that today's Wi-Fi could evolve to a third national broadband infrastructure. Perhaps the most straightforward path would follow successful development of citywide Wi-Fi networks such as those under way in Philadelphia, San Francisco, and other U.S. municipalities (see our Neutral Net scenario in Appendix A). This path might well include one or more large commercial firms (e.g., Microsoft, Google, or Yahoo!) that would provide the backbone fiber-network-linking municipal Wi-Fi as well as manage some of the local systems. A third infrastructure based on municipally-owned Wi-Fi would be likely, in our view, to embrace net neutrality and open-access concepts, although that is not an inevitable outcome. Moreover, incumbent telco and cable firms strongly oppose government-owned Wi-Fi or other broadband systems and have lobbied heavily, and often successfully, to restrict their development (see our TCNOtopia scenario in Appendix A).

A second path for Wi-Fi development would be to create a ubiquitous, open access, decentralized wireless network from the voluntary interconnection of hundreds of thousands, or millions, of existing home, commercial, and public Wi-Fi hot spots. FON, a Spanish company, has an ambitious plan to provide inexpensive wireless routers to individuals and organizations who agree to share their Wi-Fi connections in return for free roaming on other Wi-Fi systems.[23] FON's business plan relies on the introduction of new 3G mobile phones such as the Nokia E series that automatically switch to Wi-Fi whenever an open Wi-Fi signal becomes available. So far FON has focused its initial efforts largely on countries outside the United States where broadband Internet providers allow customers to share their Wi-Fi connections. While U.S. DSL and cable broadband providers generally forbid such practices, in 2007 Time Warner Cable signed a sharing agreement with FON, and AT&T began offering wireless service with the Apple iPhone that can obtain data (but, as of this writing, not voice) through available Wi-Fi networks. When, or whether, other U.S. cellular and broadband providers will enable interconnection with local Wi-Fi remains to be seen.

A third possibility, albeit futuristic, is for decentralized wireless-mesh networks to evolve from technological advances, such as those depicted in our AutoMata scenario in which cars and other vehicles serve as mobile wireless hubs. This mobile network could then interconnect with local Wi-Fi systems to spur the evolution of open, ubiquitous wireless networks in the United States and other countries.

Each of these evolutionary paths could include the use of fixed wireless technologies, such as WiMAX, to extend the range and capabilities of local wireless networks. In our view, they also would require congressional action to allocate additional unlicensed spectrum for expanded Wi-Fi and other wireless services, as has been suggested for currently unused portions of the broadcast television spectrum.[24] This could at least double the roughly 110 MHz of spectrum available for Wi-Fi in the most successful unlicensed bands below 3 GHz and go a long way toward spurring development of an alternative U.S. broadband infrastructure.

Municipally-owned Fiber Networks

Beyond Wi-Fi networks, some local governments see a municipal role in building high-capacity optical fiber networks to carry Internet traffic. The most advanced such network today is the Utah Telecommunication Open Infrastructure Agency (UTOPIA) serving 14 cities in northeastern Utah.[25] The first phase of the UTOPIA fiber network, completed in February 2006, has been funded though sales of $85 million of municipal bonds.

ISPs such as Mstar now offer phone, television, and Internet services over the UTOPIA fiber network. Residential subscribers pay $44 per month for symmetrical 15 Mbps broadband, and business customers can purchase 30 Mbps symmetrical service for $150 per month. These data rates are ten times what most asymmetrical DSL services in the United States provide at substantially higher cost per Mbps. Technically, UTOPIA's active-Ethernet network makes it easier to offer symmetric bandwidth to customers and simpler interfaces to ISPs and content providers than is possible with the passive optical networks (PONs) used by Verizon's FiOS and AT&T's Lightwave services. Although initial capital costs are somewhat higher for active-Ethernet networks than for PONs, UTOPIA's financing with twenty-year municipal bonds allows it to offer higher-speed services at more favorable rates.

UTOPIA faced many political challenges in its early years. In 2000, AT&T (then in the cable business) sought state legislation to keep local governments from competing with private networks. A last-minute amendment to the Utah Municipal Cable Television and Public Telecommunications Services Act of 2001 exempted municipal networks that did not sell retail services to customers. UTOPIA thus offers network capacity only on a wholesale basis. Mstar is currently the sole provider of voice, video, and Internet services; but UTOPIA expects to attract other providers as the system expands. Salt Lake City, Utah's largest municipality, and three other founding cities withdrew from UTOPIA in 2003–2004 after strong lobbying by Qwest, the incumbent telephone carrier.

UTOPIA shows what is technically and economically feasible today with municipally owned broadband, and its wholesale-only model provides a clear path toward competitive broadband services that can benefit its customers in northeastern Utah. Whether other U.S. cities will follow UTOPIA's example will depend more on politics than on technical or economic factors.

User-owned access

There are now numerous examples in Canada, and a few in Europe and the United States, of universities, schools, businesses, and public agencies financing direct links (usually fiber) from their internal broadband networks to points of presence (POPs) on the Internet where they can receive competitive services. This approach effectively bypasses local access monopolies or duopolies and inherently supports net neutrality.[26]

Besides paying for the initial capital cost, users must also arrange for normal operation and maintenance of their access links. This has proven feasible in Canada, where the concept of user-owned access is well advanced. At present, users must be sufficiently large and sophisticated to negotiate favorable arrangements for initial construction and subsequent operations. Institutions that already operate their own local area networks are thus the primary initial customers for user-owned access; but the concept can be extended to apartments, co-ops and condominiums, housing developments, and (eventually) individual homes.[27]

Opposition from incumbent telco and cable operators can be expected; but again, Canada provides successful examples of how such opposition can be overcome. In fact, incumbents would be likely to win the majority of user-financed construction and maintenance contracts if they embraced the concept. Firms that have extensive high-speed fiber networks, such as Level 3 and Google, as well as other large Internet content providers, might be interested in encouraging user-owned access as an alternative to the present duopoly. What is needed now, however, are some successful demonstrations of user-owned access at sufficient scale to show that the concept is viable for big broadband expansion in the United States.

Conclusion

Although net neutrality is currently at the center of debate over the future evolution of broadband communications in the United States, legislators and other policy makers need to consider it within a broader perspective. Like other countries, the United States is in the midst of transition from mini broadband, represented by DSL and cable modem, to big broadband that provides much higher speed, lower latency, and more symmetric bandwidth primarily over fiber optic and wireless links.[28] The precise path to big broadband is still unclear and may involve a number of twists and turns, but it is important that both public- and private-sector stakeholders keep this goal in mind in making near-term as well as long-term decisions.

Current U.S. policy favors broadband competition between telephone and cable companies, each of which owns and operates its own infrastructure, and such facilities-based competition is generally good for broadband customers and providers of online content and applications. At present, however, both telco and cable network operators have substantial market power, which permits them to limit the consumer benefits from TCNO competition. As a consequence, we advocate policies that encourage investment in independent broadband infrastructure facilities or remove barriers to their development. One example would be to remove restrictions on local government ownership of broadband infrastructure (wired or wireless), while encouraging them to provide competitive service offerings on these facilities (as exemplified by the UTOPIA model in Utah). A second example would be to eliminate government or TCNO-imposed barriers to user-owned broadband access facilities and, perhaps, to subsidize a few pilot projects to test scalability. Another important step would be to allocate more usable spectrum for Wi-Fi, WiMAX, and other broadband wireless applications. In each case, the overall benefits would come not just from the availability of an additional broadband alternative, but from the salutary effects this would have on telco and cable service offerings and pricing.[29]

Calls for net neutrality are a response to growing market power of the TCNO duopoly during the transition to big broadband. But because broadband is a moving target—20 Mbps may well be considered mini broadband within a very short time—it is difficult to legislate or adopt regulatory rules that will both be enforceable and remain relevant.[30] An approach set out in the March 2006 "Annenberg Center Principles for Network Neutrality" (see box 4.4) focuses on "light touch" regulation, emphasizing general principles of competition policy where network operators have significant market power, rather than detailed rules for all broadband providers. In any event, network neutrality may be only the first topic in an ongoing public debate over how to make U.S. broadband competitive, ubiquitous and affordable, as well as a continuing source of technical and social innovation.

Appendix: Three scenarios for U.S. broadband access evolution

1 Neutral Net

It's 2017. The U.S. government runs the national communication GRID (Government-Run Information Distributor), comprised of the country's fiber optics, cables, and radio links. Access to the GRID is open to all, on an equal

Box 4.4

Annenberg Center Principles for Network Neutrality (March 2006).

The goal of the Annenberg Center Principles for Network Neutrality is to provide a simple, clear set of guidelines addressing the public Internet markets for broadband access.

1. *Operators and Customers Both Should Win* It is important to encourage network infrastructure investment by enabling operators to benefit from their investments. It also is important to ensure that customers have the option of unrestricted access to services and content on the global public Internet.

2. *Light Touch Regulation* Any regulation should be defined and administered on a nationally uniform basis with a light touch. Regulations should be aimed primarily at markets in which it has been demonstrated that operators possess significant market power. The emphasis should be on prompt enforcement of general principles of competition policy, not detailed regulation of conduct in telecommunications markets.

3. *Basic Access Broadband* Broadband network operators should provide "Basic Access Broadband," a meaningful, neutral Internet connectivity service.[a] Beyond providing this level of service, operators would be free to determine all service parameters, including performance, pricing, and the prioritization of third party traffic.

4. *Transparency* Customers should receive clear, understandable terms and conditions of service explaining how any network operator, Internet service provider, or Internet content provider will use their personal information and prioritize or otherwise control content that reaches them.

5. *Encouraging Competitive Entry* Government policy should encourage competitive entry and technological innovation in broadband access markets in order to help achieve effective network competition and make available high speed Internet access to the largest number of customers.

---

a. Network operators providing basic access should not insert themselves in the traffic stream by blocking or degrading traffic. Traffic should be carried regardless of content or destination, and operators should not give preferential treatment to their own or affiliated content in the basic access service. The specific parameters (speed and latency) of this service will be reviewed on a quadrennial basis. Current thinking is that speeds of 1.25+ Mb/s downstream and less upstream would be acceptable at this time, moving to increasingly symmetric bandwidth at higher speeds in the future.

basis, for any application and any content. Most of the population now creates and shares media of all kinds—what their productions lack in polish and sophistication, they make up in imagination.

Thanks to the government-run GRID, there no longer is a divide between urban and rural areas. The open access GRID has ushered in the era of micropolitics: every conceivable constituency can propose any initiative at any time and set up a virtual debate space and e-voting mechanisms.

Neutral Net was set in motion in 1983, when the FCC forced the local phone companies to let all enhanced service providers use their wires for free. Within a few years, thousands of ISPs jumped at the chance to offer new services without the need to invest in costly networks. With the release of the Mosaic Internet browser in 1993, a new mass medium was born. Soon after, in 1995, DSL and cable modems turned the old phone and cable television networks into broadband always-on information networks.

During the next ten years, a multitude of innovators built upon the open Internet to offer new communication services that radically transformed people's ability to create, share and access information.

In our scenario, in September 2010 the U.S. Congress decided it essential to preserve the Internet's openness. Strict rules forbid all network owners, telephone, cellular, and cable alike, to discriminate among users. They are not allowed to favor any traffic, nor to charge different fees for different users or different applications.

Anybody can now provide any communication service over the carriers' networks. Wal-Mart introduces low-cost "WAL-Media": their branded combination of wired and wireless Internet access, voice and text communication, and film and video distribution.

In the next few years, amateur production of content explodes. YouTube and MySpace garner audiences that far surpass those of traditional television channels. Blogs have now replaced newspapers as most people's primary source of news. The Net supports a vibrant public sphere in which all constituencies find a voice, a virtual town hall, and viral tools to mobilize voters and make their voices heard.

To sort through this massive amount of news, debates, games, music, video, and films, users rely on each other. Social filters, recommendation engines, and distributed online marketplaces allow them to find, discover, rank, and select materials that match their passions.

Every device on the network is a server, whether in homes, public places, small businesses, or civic organizations. They support P2P communication tools; distribute user-produced stories, songs, and videos; and host collaborative spaces that bring together families, workgroups, clubs, churches, or citizens.

A growing number of cities build their own Wi-Fi and fiber networks to foster greater civic Internet use. However, funding for professionally-produced premium content starts to decline, partly because it is impossible to guarantee the network performance that would allow optimum delivery of that content and partly because P2P distribution of pirated content proliferates (it is hard to maintain control over intellectual property now that a multitude of service providers operate over the networks).

By 2012, network owners are unable to raise funds to upgrade their networks. Verizon discontinues FiOS, and AT&T abandons project Lightspeed. Cellular networks never fully upgrade to 3G. The network owners decide to become pure bit carriers, scale down their production and programming operations, and concentrate instead on cutting their costs down to a minimum, retaining only skeleton maintenance crews.

Meanwhile, although content from millions of amateur sources is now available, Hollywood loses its preeminence as the world's main center of content production. Instead, big-budget entertainment is now produced in countries like China, France, and India, where the network owners keep tight control over who distributes what and can thus guarantee protection of their intellectual property.

By 2014, investment in the U.S. network infrastructure has fallen so low that its derelict state resembles that of the nation's bridges and roads. To ward off catastrophic failure, the U.S. government takes over all communication networks, consolidating them into the GRID. A new tax on advertising is created to fund the GRID.

By 2017, the GRID provides uniform Net access throughout the U.S. territory. The nation ranks a weak twenty-ninth in the OECD's assessment of broadband performance, but a dynamic community of users constantly invents new ways to squeeze extra bits out of the country's infrastructure.

U.S. elites are dissatisfied with the poor performance of the national GRID. They live in teleparks, the new gated communities, which tend to congregate in border cities and ports, where they get easy access to foreign network headends and submarine high-capacity fiber.

2 TCNOtopia

In the year 2017, two huge TCNOs control broadband Internet access throughout the United States. Each TCNO has its own content affiliates who provide online entertainment, sports, games, and information to the consuming public. Their operational motto is "we create, you enjoy."

The path to TCNOtopia began in 1969, when the first bits sped across a new computer network funded by the U.S. Department of Defense. Soon the elements of what would become the Internet were in place: an open architecture where users innovate at the edges of the network and E2E communications with no gatekeeper inside the network core, all of it riding on top of the nation's phone network, providing little compensation to the telcos who had built that infrastructure. In fact, the Internet stands in sharp contrast to telephone and cable visions, which place intelligence, control, and innovation inside the network.

By 2007, the TCNOs provide more than 96% of residential broadband connections. But most of the real profits are made by firms who use the TCNO networks, such as Microsoft, Amazon.com, Google, Yahoo!, eBay, and Disney. Verizon and AT&T fight back with Internet television, offering hundreds of channels and thousands of hours of on-demand programs. Like the cable companies, they want to choose the content they deliver over their broadband pipes and not simply act as common carriers. AT&T's CEO declares that Internet content providers will have to pay extra for fast broadband delivery.

In reaction, content providers join with consumer groups to persuade Congress to preserve network neutrality. But they get a chilly reception in Washington. Instead, Congress gives telcos authority to freely offer Internet programming and decide what traffic gets priority within their network.

2010: Based on the early success of Wi-Fi in Philadelphia and San Francisco, Google launches broadband wireless nationwide in partnership with local municipalities.

Verizon and AT&T, followed by the cable operators, offer contracts to Sony, Fox, Disney, and others for fast-lane Internet delivery of their online games, movies, video, and other content. Those who choose not to pay must accept standard delivery. This slow lane is where user experimentation is allowed, the only option for user-run servers, and P2P and other applications unaffiliated with the carriers. To enforce the separation, the TCNOs now scan all data packets. Customer contracts authorize carriers to screen for viruses, spam, copyright violations, and content of interest to government agencies. These contracts also limit the bits users can upload without paying substantially higher fees.

2011: Most large content providers are enthusiastic about fast-lane delivery. They can now charge higher fees for premium media experiences. But some, like Google and Microsoft, mount court challenges to packet scanning and prioritization as violations of users' privacy rights and of network operators' obligations to provide common carrier services.

2012: Flush with cash from content providers, TCNOs accelerate investment in fiber infrastructure and in-network innovations to achieve high performance. Dozens of new services, such as online multiplayer sports and games, become wildly popular. With full control over individual data streams, the carriers can craft compelling multimedia experiences for their customers. TCNO interface equipment in the home also optimizes the user experience, while preventing unauthorized copying of content or the bypassing of advertising messages.

Meanwhile, Google's broadband wireless buildout has achieved initial success with four million subscribers in twenty-eight cities. However, security and reliability concerns arise after hacker attacks disable some fifteen thousand wireless-enabled computers in Chicago and Los Angeles. The TCNOs effectively use this security failure in their broadband marketing campaigns.

2014: The U.S. Supreme Court rules in favor of the TCNOs' right to scan data packets and prioritize Internet traffic. The decision cites the need to ensure network reliability and protect customers from hacker-induced harm.

2016: The merger of Comcast and Time Warner creates a behemoth controlling 90 percent of the U.S. cable market and 60 percent of all broadband connections.

After reporting billion dollar losses, Googlezon (formed by the recent merger of Google and Amazon.com)[31] abandons its municipal wireless partnerships. Some cities vow to keep their networks on the air, but it appears an uphill struggle against the dominance of TCNO broadband.

2017: Determining that only increased scale can compete effectively with Comcast Time Warner, the Justice Department approves the merger of AT&T and Verizon. The broadband duopoly has no serious rivals. It has brought affordable broadband to 85 percent of U.S. households, who love the network innovations that protect against spam and viruses, the e-sports leagues, and the high-definition entertainment they receive from TCNO content affiliates.

Political expression online is encouraged within the established political structure—primarily through the two dominant national parties that have negotiated fast-lane delivery for their candidates and issue messages. Other political organizations and civic groups must negotiate ad hoc arrangements, and few have the financial resources to assure fast lane delivery of their messages.

Still, some academics, artists, and other dissidents bemoan the loss of amateur content production and collaborative activity that flowed over the Internet in the early twenty-first century. TCNO restrictions have virtually eliminated P2P communication among residential broadband users for content distribution, collaborative work, or social and political organization. Online distribution and collaboration are channeled through TCNO-controlled servers and routers. As a consequence, Internet content and applications now conform closely to established consumer tastes and traditional values. It is nearly impossible for an innovator that is unaffiliated with the TCNOs to gain a sizable Internet audience in the United States

But perhaps another eBay, Napster, Yahoo!, Amazon.com or Google is ready to emerge out of the competitive chaos in India or Brazil.

3 AutoMata

In the year 2015, the Internet addresses the last-mile challenge using a mesh network of wireless devices named AutoMatas. The traditional telcos are now relegated to routing the backbone traffic. AutoMatas are self-organizing devices that communicate with one another when in each other's radio range. They may use access points available at hot spots for access to the telcos' wired infrastructure. AutoMatas are now ubiquitous in society and a standard feature in all new vehicles. Their widespread use contributes to augmenting the capacity of the wireless mesh network that effectively transmits the last-mile traffic.

An early example of wireless ad-hoc networks is the use of short-range radios by truck drivers to communicate road conditions and socialize while traveling. The need to exchange information and communicate while in motion was later satisfied through the use of mobile phones, although this system does not operate on an ad-hoc mesh network. The origins of the wireless ad-hoc network organized around AutoMatas dates back to 1991, when Vic Hayes of NCR Corporation in the Netherlands invents Wi-Fi for use with cashier systems. Prior to his retirement in 2003, he shapes the design of standards such as IEEE 802.11a, b, and g, becoming the "father of Wi-Fi." The devices that operate on this standard have a limited radio range in the order of tens of feet and offer bandwidths in the order of tens of megabits per seconds.

In the first years of the new millennium, a number of different events begin to lay the ground for the emergence of a wireless mesh network. Lightweight portable products with gigabytes of memory storage, such as the iPod, hit the market enabling the consumption of digital entertainment on the go. COVERGE 2001 is a first conference dedicated to convergence of automobiles and computers. Participants discuss standards for an in-vehicle multimedia network. Mesh networks emerge in neighborhoods using Wi-Fi devices based on IEEE 802.11a, b, and g, and offering Internet services to their residents.

2009: Advent of Wi-Fi tailor-made for mobile vehicles. The technological leap to enable Wi-Fi on mobile vehicles occurs in 2009, when companies introduce products based on 802.11p, also referred to as Wireless Access for the Vehicular Environment (WAVE), offering radio ranges in the order of a thousand feet with bandwidths in the order of megabits per second. Such ranges make possible meaningful exchange of information between vehicles that are in motion and base stations.

2010: The U.S. Congress makes a significant part of the analog broadcast spectrum available for unlicensed use. Groundbreaking work in interference-avoidance techniques has enabled efficient use of this spectrum. This big-broadband wireless spectrum represents the future of wireless communications. The unlikely group of successful bidders for the spectrum is composed of EarthLink, Sony, Cingular, GM, and Toyota. While EarthLink and Sony are preparing for general distribution of wireless content, the car companies are observing this from the consumer's end: soon vehicles will be capable of receiving massive quantities of wireless content reliably, and new vehicle designs in the works are incorporating these features.

2011: The year 2011 witnesses the development of AutoMata. These wireless devices, looking like iPods and weighing less than 10 ounces, store audio and video clips, implement navigation and GPS capabilities of an automobile, and detect one another for multi-player games that extend a physical world with virtual objects. AutoMatas include terabytes of storage, and several types of wireless cards such as cellular, 802.11g, 802.11p, and Bluetooth. These networking cards operate in a variety of radio ranges (from a few feet to miles) and bandwidths (tens of kilobits to hundreds of megabits per second). Numerous car manufacturers begin to combine the vehicle's telematics, navigation, and entertainment systems into AutoMatas for new models. These devices plug into an in-vehicle network to guide a driver to a destination and deliver entertainment content. Its owner may carry the device and use it as a personal digital assistant and personal audio and video entertainment unit.

2012: Hackers publish the programming interface to AutoMata, adding features such as text-to-audio and a VoIP interface. Using a Bluetooth-enabled headset, a user may listen to e-mail messages. Consumer electronic vendors are quick to adopt these novel features, publishing official versions for download by all. This grassroots effort introduces many safety features for in-vehicle use. For example, when in radio range of one another, AutoMatas exchange traffic information and hazards such as icy road conditions. An elegant holder on the front dashboard of a car provides for both recharging of the AutoMata and its view of the road in front of the automobile. A passenger may view what another vehicle's AutoMata might be recording several minutes ahead. Insurance companies start to give incentives to drivers who use voice activated AutoMatas that warn them of road hazards.

2013: Competing AutoMata-like devices reach the market, driving down prices and improving their efficiency and capabilities. In addition an active community of programmers continue to develop applications and operating systems and standards for AutoMatas. Devices are everywhere, creating an ad-hoc mesh network that begins to produce and route growing amounts of Internet traffic. In particular, the planned pre-staging of popular content to make it accessible to the mesh network is effectively reducing the last-mile Internet traffic. Users begin to switch from home ISP subscriptions to maintaining their AutoMata connectivity, much like mobile phone usage affected standard telephone lines. Telcos point to the growing traffic created by the mesh network, and its lack of security, to insist on the need to manage traffic on the Internet backbone to justify the investment in increased capacity. There is a surge in spam and identity theft due to the increasing Internet traffic on the mesh network. Soon telcos develop enhanced certification methods for personal messages and improved cryptography techniques for security introduced by academics and other users of the network.

2014: GM and Ford introduce software for AutoMata, enabling cars from 2014 onward to drive without human assistance on certain freeway stretches. A valid driver's license holder must sit in the driver's seat and agree to the risks of using AutoMata. When an intelligent road stretch is encountered, AutoMata signals its driver that it may assume responsibility for driving the car. These devices communicate with one another to share information about icy road conditions, accidents such as chemical spills, and other emergency situations. Devices minimize accidents by slowing down and stopping when too close to one another.

There is a steady reduction in the use of mobile phones, the mesh network provides better coverage in most urban settings than the nonintegrated cellular networks due to the ubiquitous presence of the AutoMata. Mobile phone calls are now mostly conducted with the AutoMata using VoIP.

2015: An AutoMata is now the size of a company pin that one wears on a business suit. It records and stores hundreds of hours of phone and live conversation, along with hours of video recordings. Its battery works for days of usage. For their display, AutoMatas utilize TVs in a living room, a laptop's display at work, a mobile phone's display on the road, and a car's navigation display or fold-down screen. New and inexpensive displays start to appear in automobiles and coffeehouses. Menus at high-end restaurants turn into a display and a keyboard for AutoMatas.

A year after the introduction of automatic driving by AutoMatas, there is a significant reduction in the number of accidents, and many insurance companies give incentives to those who employ AutoMata to drive. In collaboration with dozens of insurance companies, the U.S. Department of Transportation deploys Internet wireless hubs on most highway stretches in the country. Its primary purpose is to increase road safety and enhance emergency response.

This new infrastructure helps create an alternate Internet backbone to the mired telco network and its traffic-management practices. There is now an alternative backbone for delivery of delayed modes of communication, alleviating the load on the telco's wired backbone infrastructure.

2017: In cities such as Los Angeles a growing number of people are driven by their AutoMata from place to place. A vehicle then has become an extension of both the office and home. Special content is released for in-vehicle entertainment systems, capitalizing on its close proximity to immerse passengers in an experience.

Bollywood of India finally supersedes Hollywood in revenues by generating content tailor-made for in-vehicle use. Widespread amateur production and improved social filters have enabled a broad dissemination of diverse content and ideas, effectively "fattening the long tail" for entertainment, commerce, and political movements. Every idea is capable of finding its audience, however, due to the large number of possibilities available, advertising on the Internet and on hardware is still able to shape public opinion. Much like important content producers such as Disney, powerful interest groups such as the NRA and Texans for a Democratic Majority promote their message through product placement on antivirus and antispam software and new equipment. In addition, content can be prioritized by pre-placing content through the access points to the mesh network for a fee.

Widespread use of AutoMata has lead to a decline of home ISPs; local Internet traffic is routed through the mesh network, and traditional telcos are relegated to routing part of the backbone traffic. Bandwidth and latency on the mesh network is affected by the number of AutoMata present.

Notes

1. Jeffrey Hart, François Bar, and Robert Reed, "The Building of the Internet: Implications for the Future of Broadband Networks," *Telecommunications Policy* 16, no. 8 (1992): 666–689.

2. Jerome H Saltzer, David P. Reed, and David D. Clark, "End-to-end arguments in system design," *ACM Transactions on Computer Systems* 2, no. 4 (1984): 277–288.

3. François Bar, Stephen Cohen, Peter Cowhey, Brad DeLong, Michael Kleeman, and John Zysman, "Access and Innovation Policy for the Third-Generation Internet," *Telecommunications Policy* 24, no. 6/7 (2000): 489–518.

4. Carterfone argued that "where a carrier has monopoly control over essential facilities we will not condone any policy or practice whereby such carrier would discriminate in favor of an affiliated carrier or show favoritism among competitors." See Federal Communications Commission, 29 F.C.C.2d 870, 1971, para 157; See, also, "In the Matter of Use Of The Carterfone Device In Message Toll Telephone Service," Docket

No. 16942, 13 F.C.C.2d 420, June 26, 1968; MCI v. FCC (Execunet I), 561 F.2d 365 (D.D.C. 1977), cert. denied, 434 U.S. 1041 (1978); MCI v. FCC (Execunet II), 580 F.2d 590 (D.D.C.), cert. denied 439 U.S. 980 (1978); Computer I, 28 F.C.C.2d 267 (1971); Computer II, 77 F.C.C.2d 384 (1980); Computer III *Notice of Proposed Rulemaking,* F.C.C. 85–397 (Aug. 16, 1985).

5. Organisation for Economic Co-operation and Development, "OECD Broadband Statistics to December 2006," http://www.oecd.org/document/7/0,3343,en_2649_34223_38446855_1_1_1_1,00.html.

6. Anindya Chaudhuri and Kenneth Flamm, "An Analysis of the Determinants of Broadband Access," *Telecommunications Policy* 29, no. 9/10 (2005): 731–755.

7. One exception noted in OECD, "OECD Broadband Statistics," and discussed later in this chapter is broadband over fiber-optic facilities financed by local governments in northeastern Utah.

8. Scott Wallsten, "Broadband and Unbundling Regulations in OECD Countries," (working paper, AEI-Brookings Joint Center For Regulatory Studies, Working Paper 06–16, June 2006), http://www.aei-brookings.org/publications/abstract.php?pid=1084.

9. However, Wallsten, "Broadband and Unbundling Regulations," (AEI-Brookings Joint Center Working Paper No. 06-16, June 2006), Social Science Research Network, http://ssrn.com/abstract=906865, disputes the importance of unbundling, concluding that it has had no significant impact on broadband penetration across OECD countries.

10. For a representative and well-articulated summary of these views, see Public Knowledge, Network Neutrality Overview, available at http://www.publicknowledge.org/issues/network-neutrality.

11. See Ithiel de Sola Pool, *Technologies of Freedom,* (Cambridge, MA: Harvard University Press, 1983); François Bar and Christian Sandvig, "Rules from Truth: Post-Convergence Policy for Access," (paper presented at the 28th Annual Telecommunications Policy Research Conference, Arlington VA, September 23–25, 2000).

12. Organisation for Economic Co-Operation and Development, Working Party on Telecommunication and Information Services Policies, Committee For Information, Computer And Communications Policy, Directorate For Science, Technology and Industry, "Multiple Play: Pricing And Policy Trends," April 7, 2006, http://www.oecd.org/dataoecd/47/32/36546318.pdf.

13. Declan McCullagh, "Telco agrees to stop blocking VoIP calls," *CNET.com,* March 3, 2005, http://news.com.com/2102-7352_3-5598633.html.

14. Tim O'Reilly, "What Is Web 2.0: Design Patterns and Business Models for the Next Generation of Software," *O'Reilly Network,* September 30, 2005, http://www.oreillynet.com/lpt/a/6228.

15. Wikipedia contributors, "Comparison of latency and throughput," Wikipedia, The Free Encyclopedia, http://en.wikipedia.org/w/index.php?title=Comparison_of_latency_and_throughput&oldid=95424213.

16. Shahram Ghandeharizadeh, Bhaskar Krishnamachari, and Shanshan Song, "Placement of Continuous Media in Wireless Peer-to-Peer Networks," *IEEE Transactions on Multimedia* 6, no 4, (April 2004); International Telecommunication Union, "The Internet of Things," (ITU Internet Reports, 2005), http://www.itu.int/osg/spu/publications/internetofthings.

17. International Telecommunication Union, *Internet of Things.*

18. By 2010, Verizon expects to spend $20 billion to expand its FiOS fiber-to-the-home network (which will be similar to the fiber networks in Japan and Korea) to 16.1 million households; AT&T will invest $7 billion to serve 18.3 million households with its less costly but less capable fiber-to-the-neighborhood approach. Craig Moffett, "The 'Dumb Pipe' Paradox (Part II) Patchwork Pipes," (corporate report, Sanford C. Bernstein & Co., 2006).

19. David D. Clark and Marjory S. Blumenthal, "Rethinking the Design of the Internet: The End-to-End arguments vs. the Brave New World," *ACM Transactions on Internet Technology* 1, no 1, (2001): 70–109.

20. ibid.

21. For references, see David Reed, "Open Spectrum Resource Page," http://www.reed.com/OpenSpectrum/.

22. Charles Jackson, Raymond Pickholtz, and Dale Hatfield, "Spread Spectrum Is Good—But It Does Not Obsolete NBC vs. U.S.!" *Federal Communications Law Journal* 58, no 2 (April 2006), 245–264. http://www.law.indiana.edu/fclj/pubs/v58no2.html.

23. Eric Auchard, "Bargain Routers to Boost Wireless," *Australian IT,* June 25, 2006, http://www.australianit.news.com.au/story/0,24897,19590235-15350,00.html (we should note that using FON may violate a customer's contract when carriers prohibit connection sharing. To date however, this hasn't emerged as a significant problem).

24. Pierre de Vries, "Populating the Vacant Channels: The Case for Allocating Unused Spectrum in the Digital TV Bands to Unlicensed Use for Broadband and Wireless Innovation," (Working Paper 14, New America Foundation, August 2006), http://www.newamerica.net/files/WorkingPaper14.DTVWhiteSpace.deVries.pdf.

25. Steven Cherry, "A Broadband Utopia [Fast Broadband Connectivity]," *Spectrum, IEEE* 43, no. 5 (2006), 48–54.

26. IEEE 802.17 Resilient Packet Ring Working Group, "The Case for Deploying 'Big Broadband' Through Open Advanced Fiber Networks (AFNs)," (executive summary, 2004), http://www.ieee802.org/17/email/pdf00058.pdf.

27. Bill St. Arnaud, "A Business Strategy To Avoid The Two Tier Internet," (presentation, CANARIE, Inc., 2006), http://www.canarie.ca/canet4/library/customer/Last_Mile_Customer_Owned_Networks.pdf.

28. Again, we recognize that other technologies such as satellite and broadband over power lines will participate in this transition, but, in our view, probably as relatively minor niche players.

29. U.S. court decisions and regulatory policies no longer support unbundling of telco or cable broadband access facilities to allow competition in services, as contemplated under the Telecommunications Act of 1996. As described in Section 1, however, unbundling has led to effective competition in broadband services in Japan, South Korea, France, and several other OECD countries. If U.S. policies to promote facilities-based competition with the telco/cable duopoly are not successful, Congress should reconsider unbundling or other ways to achieve effective competition in broadband services.

30. Moreover, some experts contend that the current focus on network neutrality by Internet content providers is a "dangerous sideshow" distracting attention from telco efforts to dominate the Internet backbone as well as local access networks. See Gordon Cook, "Fighting the Wrong Net Neutrality War," *The Cook Report on Internet Protocol,* Cook Network Consultants, (September 2006).

31. Robin Sloan, "EPIC 2014," (2004), http://epic.makingithappen.co.uk/ols-master1.html.

Conclusion: The Meaning of Network Culture

Kazys Varnelis

Taken together, the chapters in this book point to the development of a new societal condition, spurred by the maturing of the Internet and mobile telephony. In this conclusion, I will reflect on that state—which I will call *network culture*—as a broadly historical phenomenon. Defined by the very issues that these chapters raise—the simultaneous superimposition of real and virtual space, the new participatory media, concerns about the virtues of mobilization versus deliberation in the networked public sphere, and emerging debates over the nature of access—network culture can also reveal broader societal structures, just as concepts such as modernism and postmodernism did in their day.

Although subtle, this shift in society is real and radical. During the space of a decade, the network has become the dominant cultural logic. Our economy, public sphere, culture, and even our subjectivity are mutating rapidly and show little evidence of slowing down the pace of their evolution. One morning we note with interest that our favorite newspaper has established a Web site. Another day we decide to stop buying the paper and just read the site. Then we start reading it on a mobile Internet platform, or listening to a podcast of our favorite column, while riding a train. Or perhaps we dispense with official news entirely in favor of a collection of blogs and amateur content. When we buy our first mobile phone we are unaware of how profoundly it will alter our lives. But soon we forget shopping lists in favor of the immediacy of calling home from the store to see what's in the refrigerator. We stop scheduling dinner plans with friends long in advance when we can instead coordinate them en route to a particular neighborhood. When we move away from intimate friends or family, we no longer have to lose touch. Even going away to college has a new meaning when children can call their parents just to say hi as they cross campus on their way to class. Our chance visit to a friend's Web site shocks us with the news that he has passed away suddenly, the daily updates of his battle against sudden illness cut short. Individually, such everyday narratives of how technology reshapes our lives are minor. Collectively, they are deeply transformative.

Network culture extends the information age of digital computing.[1] But it is also markedly unlike the PC-centered time that culminated in the 1990s. Indeed, in many ways we are more distant from the the era of PC-centered computing than it was from the time of centralized, mainframe-based computation. To understand this shift, we can usefully employ Charlie Gere's insightful discussion of computation in *Digital Culture.* In Gere's analysis, much as in the methodology that we've adopted throughout this book, the digital is a socioeconomic phenomenon as much as a technology. Digital culture, he observes, is fundamentally based on a process of abstraction that reduces complex wholes into more elementary units. Tracing this process of abstraction to the invention of the typewriter, Gere identifies digitization as a key process of capitalism. By separating the physical nature of commodities from their representations, digitization enables capital to circulate more freely and rapidly. In this ability to turn everything into quantifiable, interchangeable data, digital culture is universalizing. Gere cites the universal Turing machine—a hypothetical computer first described by Alan Turing in 1936, capable of being configured to do any task—as the model for not only the digital computer but also for that universalizing aspect of digital culture.[2]

But today connection is more important than division. In contrast to digital culture, under network culture information is less the product of discrete processing units than of the outcome of the networked relations between them, of links between people, between machines, and between machines and people.

Perhaps the best way to illuminate the difference between digital culture and network culture is to contrast their physical sites. The digital era is marked by the desktop microcomputer, displaying information through a heavy CRT monitor, connected to the network via dial-up modem or perhaps through a high-latency first-generation broadband connection. In our own day, there is no such dominant site. The desktop machine is increasingly relegated to high-end applications such as graphic rendering or cinema-quality video editing or is employed for specific, location-bound functions (at reception desks, to contain secure data, as point-of-sale terminals, in school labs, and so on) while the portable notebook or laptop has taken over as the most popular computing platform. But the laptop can be used anywhere: in the office, at school, in bed, in a hotel, in a café, or on the train or plane. Not only are networks an order of magnitude faster than they were in the dial-up days of the PC, but Wi-Fi makes them easily accessible in many locations. Smart phones such as the Blackberry, Treo, and the iPhone complement the laptop, bringing connectivity and processing power to places that even laptops can't easily inhabit, such as streets, subways, or automobiles. But such ultraportable devices are also increasingly competing with the computer, taking over functions that were once in the universal device's purview.[3] What unites these machines is their mobility and their interconnectivity, necessary to make them more ubiquitous companions in our lives and key interfaces to global telecommunications networks. In a prosaic sense, the Turing machine is already a reality, but it doesn't take the form of one machine, it takes the form of many. With minor exceptions, the laptop, smart phone, cable TV set-top box, game console, wireless router, iPod, iPhone, and Mars rover, are the same device, becoming specific only in their interfaces, their mechanisms for input and output, for sensing and acting upon the world. Instead, the new technological grail for industry is a universal, converged network, capable of distributing audio, video, Internet, voice, text chat, and any other conceivable networking task efficiently.

Increasingly, the immaterial production of information and its distribution through the network is the dominant organizational principle for the global economy. To be clear, we are far from the world of immaterial production. We manufacture physical things, even if increasingly that manufacturing happens in the developing world. Moreover, the ease of obtaining goods manufactured far away is due to the physical network of global logistics. Sending production offshore—itself a consequence of new network flows—may put it out of sight, but doesn't reduce its impact on the Earth's ecosystem. And, beyond global warming, even in the developed world there are consequences: Silicon Valley contains more EPA (U.S. Environmental Protection Agency) Superfund sites than any other county in the nation.[4] But as Saskia Sassen and Manuel Castells have concluded, regardless of our continued dependency on the physical, the production of information and the transmission of that information on networks are the key organizing factors in the world economy today. Although other ages have had their networks, ours is the first modern age in which the network is the dominant organizational paradigm, supplanting centralized hierarchies.[5] The ensuing condition, as Castells suggests in *The Rise of the Network Society,* is the product of a series of changes: the change in capital in which transnational corporations turn to networks for flexibility and global management, production, and trade; the change in individual behavior, in which networks have become a prime tool for individuals seeking freedom and communication with others who share their interests, desires, and hopes; and the change in technology, in which people worldwide have rapidly adopted digital technology and new forms of telecommunication in everyday life.[6]

As we might expect, the network goes even further, extending deeply into the domain of culture. In the same way that network culture builds on digital culture, it builds on the culture of postmodernism outlined by Fredric Jameson in his seminal essay "Postmodernism, or the Cultural Logic of Late Capitalism," first written in 1983 and later elaborated upon in a book of the same title. For Jameson, postmodernism was not merely a stylistic movement but rather a broad cultural determinant stemming from a fundamental shift to the socioeconomic phase of history that economist Ernest Mandel called "late capitalism." Both Mandel and Jameson concluded that society had been thoroughly colonized by capital under late capitalism, and any remaining precapitalist forms of life had been absorbed.[7] Mandel situated late capitalism within a historical model of long-wave Kondratieff cycles. These economic cycles, comprised of twenty-five years of growth followed by twenty-five years of stagnation, provided a compelling model of economic history following a certain rhythm: fifty years of Industrial Revolution and handcrafted steam engines culminating in the political crises of 1848, fifty years of machined steam engines lasting until the 1890s, electric and internal combustion engines underwriting the great modern moment that culminated in World War II, and the birth of electronics marking the late capitalism of the postwar era.[8]

If digital culture flourished during late capitalism, then it should not be surprising that Jameson observed in that period that everything became interchangeable, quantified, and exchangeable for money or other items. With the gold standard done away with, capital is valued purely for its own sake, no longer a stand-in for something else but, rather simply, pure value. The result is the disappearance of any exterior to capital and with it the elimination of any place from which to critique or observe capital. As a consequence, postmodern culture loses all meaning and any existential ground or deeper meaning. Depth, and with it emotion, vanished, to be replaced by surface effects and intensities. In this condition, even alienation was no longer possible. The subject became schizophrenic, lost in the hyperspace of late capital.

As capital colonized art under late capitalism, Jameson suggested, even art lost its capacity to be a form of resistance. The result was a cross-contamination as art became not just an industry but an investment market, and while artists, fascinated by the market, began to freely intermingle high and low. With the art market calling for easy reproducibility and marketing, and with authenticity no longer a viable place of resistance, some artists began to play with simulation and reproduction. Others, finding themselves unable to reflect directly on the condition of late capital but still wanting to comment upon it, turned to allegory, foregrounding its fragmentary and incomplete nature.

History, too, lost its meaning and purpose, both in culture and in academia. In the former, history was instead recapitulated as nostalgia, thoroughly exchangeable and made popular in the obsession with antiques, as well as through retro films such as *Chinatown, American Graffiti, Grease,* or *Animal House.* In academia, a spatialized theory replaced historical means of explanation as a means of analysis.

Modernism's obsession with its place in history was inverted by postmodernism, which, as Jameson points out, was marked by a waning of historicity, a general historical amnesia. But if postmodernism undid its ties to history to an even greater extent than modernism, it still grounded itself in history, both in name—which referred to its historical succession of the prior movement—and in its delight in poaching from both the premodern past and the more historically distant periods of modernism itself (e.g., Art Nouveau, Russian revolutionary art, Expressionism, Dada).

Today, network culture succeeds postmodernism. It does so in a more subtle way. No new "ism" has emerged: that would lay claim to the familiar territory of manifestos, symposia, definitive museum exhibits, and so on. Instead, network culture is a more emergent phenomenon.

Evidence that we have moved away from postmodernism can be found in economic cycles. If late capitalism is still the economic regime of our day, it will be the longest lasting of all the Kondratieff cycles. Assuming the Kondratieff cycles are accurate, Jameson's theorization would come in a downswing on the cycle that began after World War II. Indeed, given the protracted economy downturn of postfordist restructuring during the 1970s and 1980s, this seems entirely reasonable. A critical break took place in 1989 with the fall of the Soviet Union and the integration of China into the world market, instantiating the "new" world order of globalization. In turn, the commercialization of the Internet during the early 1990s set the stage for massive investment in the crucial new technology necessary for the new, fresh cycle. New Kondratieff cycles are marked by spectacular booms, so the delirious dot-com boom and the more docile, seemingly more sustainable, upswing of Web 2.0 would then be legible as the first and second booms of a Kondratieff cycle on the upswing. It is this second upswing, then, in which network culture can be observed as a distinct phenomenon.

Even if we are to abandon the Kondratieff cycles as overly determinist, since the turn of the twentieth century, at least, no cultural movement has lasted more than twenty-five years. It would require special dispensation to argue that we are still in the same moment as Jameson was when he first formulated his thesis.

The closest thing we have to a synthetic understanding of this era is the political theory laid out in Michael Hardt and Antonio Negri's *Empire.* In their analysis, the old world order based on the imperialist division of the globe into spheres of influence has been superseded by "Empire," a diffuse power emanating not from any one place but, rather, from the network itself. Empire's economy is immaterial; its power not only stems from the economic force of capital, but is constructed by juridical means. As nation-states fade away under globalization, to ensure mobility and flexibility of capital across borders, Empire turns to transnational governing bodies such as the United Nations to call for a universal global order. In doing so, however, Empire reinscribes existing hierarchies and, as the wars in the Middle East show, has to resort to violence. Hardt and Negri identify networked publics, which they call the multitude, as a counterforce. For them, the multitude is a swarm intelligence, able to work within Empire to demand the rights of global workers. As we have described throughout this book, the networking of individuals worldwide gives them new links and new tools with which to challenge the system, but as the chapter on politics suggests, whether networked publics can come together to make decisions democratically is still unclear.[9]

If Empire is a political theory, my goal here is to sketch out a cultural theory of this networked age. Although postmodernism anticipated many of the key innovations of network culture, our time is distinctly different.[10] In the case of art and architecture, Jameson suggests, a widespread reaction to the elitism of the modern movement and the new closeness between capital and culture led to the rise of aesthetic populism. Network culture exacerbates this condition as well, dismissing the populist *projection* of the audience's desires onto art for the *production* of art by the audience and the blurring of boundaries between media and public. If appropriation was a key aspect of postmodernism, network culture almost absentmindedly uses remix as its dominant process. A generation after photographer Sherri Levine reappropriated earlier photographs by Walker Evans, dragging images from the Internet into PowerPoint is an everyday occurrence, and it is hard to remember how radical Levine's work was in its redefinition of the Enlightenment notions of the author and originality.[11]

Art critic Nicholas Bourriaud states that this lack of regard for originality is precisely what makes art based on what he calls postproduction appropriate to network culture. Works like Levine's still relied on notions of authorship and originality for the source of their meaning. More recently, Bourriaud explains, artists like Pierre Huyghe, Douglas Gordon, or Rirkrit Tiravanija no longer question originality but rather instinctively understand artworks as objects constituted within networks, their meaning given by their position in relation to others and their use.[12] Like the DJ or the programmer, Bourriaud observes these artists "don't really 'create' anymore, they reorganize."[13]

The elements that artists choose to remix, however, tend to be contemporary.[14] The nostalgia culture so endemic to postmodernism has been undone, and the world still in the throes of modernization is long gone. Unable to periodize, network culture disregards both modern and premodern equally, as well as the interest in allegory. As T. J. Clark describes it, modernism is our antiquity, the unintelligible ruins of a vanished civilization. For Clark, like Jameson, modernism was rendered anachronistic once the process of modernization was complete.[15]

Instead of nostalgia and allegory, network culture delivers remix, shuffling together the diverse elements of present-day culture, blithely conflating high and low—if such terms can even be drawn anymore in the long tail of networked micropublics—while poaching its as-found contents from the world.[16] Correspondingly, reality increasingly dominates forms of cultural production: reality television shows are common; film documentaries such as *Supersize Me, An Inconvenient Truth,* and *Fahrenheit 911* proliferate; popular sites Web such as eBaum's World or YouTube are filled with videos that claim to be true, such as scenes of people doing incredibly stupid or dangerous things and video blogs. When fiction is deployed on Internet video sites, it poses as reality for viral marketing methods (e. g., Lonelygirl15 or Little Loca). The vision William Gibson had in *Pattern Recognition* of an exquisite movie released cut-by-cut on the Internet is replaced, instead, by low-quality clips of snarky teenagers in front of webcams or low-quality clips of actors playing snarky teenagers in front of webcams.[17]

Video games are the dominant form of fiction today. By 2004 they generated more revenue than Hollywood made in box-office receipts. If the novel simulated the internal voice of the subject, video games produce a new sort of fiction and affirm the networked self through a virtual reality in which the player can shape his or her own story. In MMORPGs such as World of Warcraft (which earns some $1 billion a year in subscription fees, compared to the $600 million earned by Hollywood's most successful product, *Titanic*) the ability to play with thousands of other individuals in immense landscapes thoroughly blurs the boundaries of reality and fiction and the boundaries of player and avatar.[18]

To be clear, the tactics of remix and the rapt fascination with reality aren't just found in GarageBand and YouTube mash-ups, they form an emerging logic in the museum and the academy as well. Art itself, long the bastion of expression, is now dominated by straightforward photography (like Andreas Gursky), and some of the most interesting work can be found in research endeavors that could easily take place in Silicon Valley rather than in the gallery (like locative media), by (sometimes carefully faked) studies of the real (like the Museum of Jurassic Technology, the Center for Land Use Interpretation, Andrea Fraser, Christoph Büchel, etc.). Other works, such as Ólafur Elíasson's ambient forms or Andrea Zittel's environments, clothing, restaurants, and High Desert Test Sites, suggest another strategy of new realism in which art becomes a background to life. Similarly, architecture has abandoned utopian projections, nostalgic laments, and critical practice alike for a fascination with the world. Arguably the world's foremost practitioner, Rem Koolhaas produces book after book, matter-of-factly announcing his fascination with shopping, the Pearl River Delta, or Lagos, Nigeria.

What of the subject in networked culture? Under modernism, for the most part, the subject is autonomous, or at least subscribes to a fantasy of autonomy, even if experiencing pressures and deformations from the simultaneity generated by that era's technologies of communication and by increasing encounters with the Other. In postmodernism, Jameson explains, these pressures couple with a final unmooring of the self from any ground as well as the undoing of any coherent temporal sequence, forcing the subject to schizophrenically fragment. With network culture, these shards of the subject take flight, disappearing into the network itself. Less an autonomous individual and more of a construct of the relations it has with others, the contemporary subject is constituted within the network. This is a development of the condition that Castells observes in *The Rise of the Network Society* when he concludes that contemporary society is driven by a fundamental division between the self and the net. To support his argument, Castells turns to Alain Touraine: "in a post-industrial society, in which cultural services have replaced material goods at the core of its production, *it is the defense of the subject, in its personality and in its culture, against the logic of apparatuses and markets, that replaces the idea of class struggle.*"[19]

But the defense of the subject has dwindled in the time since Castells and Touraine formulated their critique. Instead, it is Gilles Deleuze's "Postscript on Societies of Control" that seems more appropriate to network culture. Here Deleuze suggests that today the self is not so much constituted by any notion of identity but rather is composed of "dividuals."[20] Instead of whole individuals, we are constituted in multiple micropublics, inhabiting simultaneously overlapping telecocoons, sharing telepresence with intimates in whom we are in near-constant touch, classified into one of the sixty-four clustered demographics units described by the Claritas corporation's PRIZM system.

In network theory, a node's relationship to other networks is more important than its own uniqueness. Similarly, today we situate ourselves less as individuals and more as the product of multiple networks composed of both humans and things. This is easily demonstrated through some everyday examples. First, take the way the youth of today affirm their identities. Teens create pages on social networking sites such as MySpace and Facebook. On these pages they list their interests as a set of hyperlinked keywords directing the reader to others with similar interests. Frequently, page creators use algorithms to express (and thereby create) their identities, for example, through a Web page that, in return for responses to a set of questions, suggests what chick-flick character the respondent is.[21] At the most reductive, these algorithms take the form of simple questionnaires to be filled out and posted wholesale on one's page. Beyond making such links, posting comments about others and soliciting such comments can become an obsessive activity. Affirming one's own identity today means affirming the identity of others in a relentless potlatch. Blogs operate similarly. If they appear to be the public expression of an individual voice, in practice many blogs consist of material poached from other blogs coupled with pointers to others in the same network, for example, *trackbacks* (notifications that a blogger has posted comments about a blog post on another blogger's blog) or *blogrolls* (the long lists of blogs that frequently border blog pages). With social bookmarking services such as del.icio.us or the social music platform last.fm, even the commentary that accompanies blog posts can disappear and the user's public face turns into a pure collection of links. Engaging in telepresence by sending SMS messages to friends or calling family on a mobile phone has the same effect: the networked subject is constituted by networks both far and near, large and small. Like the artist, the networked self is an aggregator of information flows, a collection of links to others, a switching machine.

Along with this change in the self comes a new attitude toward privacy. Many blogs reconfigure the personal and the public, as individuals reveal details that had previously been considered private. The idea of locks on diaries today seems almost preposterous as individuals, particularly teenagers, discuss their most intimate—and illicit—details online.[22] Meanwhile, advances in computation and networking have made it possible to store data on individuals to a greater degree than ever imaginable. As debit cards and other technologies replace cash, our actions, be they online or out on the town, leave behind a trail of information. Corporations routinely track what Web sites individuals visit at work. In the wake of 9/11, both the United States government and others have taken to recording more and more communications traffic, even when that recording is of questionable legality. As tracking has increased, advances in data mining mean that those wishing to find information can do so more easily than ever before.

But if this degree of surveillance conjures images of George Orwell's *1984,* there has been relatively little protest. That Watergate undid Nixon seems impossible in retrospect. To some degree this is the case of what security researcher Ross Anderson calls "boiling the frog" (a frog in a pot of water doesn't notice when the temperature of the water is raised incrementally, and it boils to death).[23] Nevertheless, it also underscores the degree to which privacy is no longer important in this culture. As the subject is increasingly less sure of where the self begins and ends, the question of what should be private and what shouldn't be fades.

Under network culture, then, the waning of the subject that began under postmodernism grows ever greater. But whereas under postmodernism, being was left in a free-floating fabric of emotional intensities, today it is found in the net. The Cartesian "I think therefore I am" dissolves in favor of an affirmation of existence through the network itself, a phantom individuality that escapes into the network, much as meaning escapes into the Derridean network of *différance,* words defined by other words, significance endlessly deferred in a ceaseless play of language.[24] The division between the self and the net that Castells observed a decade ago is undone.

Nor are the networks that make up the contemporary self merely networks of people. On the contrary, they are also networks between people and things. In Bruno Latour's analysis, things are not merely objects that do our bidding but are key actors in the network. As things get smarter and smarter, they are ever more likely to make up larger parts of our "selves." An iPod is nothing less than a portable generator of affect with which we paint our environment, creating a soundtrack to life. A Blackberry or telephone constantly receiving text messages encourages its owner to submit to a constantly distracted state, a condition much lamented by many.[25]

It is in this context that networked publics form. Apart from the loss of the self, of all the changes that network culture brings us, the reconfiguration of the public sphere is likely to be the most significant, a distinction that makes our moment altogether unlike any other in three centuries. Since the Enlightenment era, the public came to be understood as a realm of politics, media, and culture, a site of display and debate open to every citizen while, in turn, the private was broadly understood as a realm of freedom, inwardness, and individuality. The public sphere was the space in which bourgeois culture and politics played out, a theater for the bourgeois citizen to play his role in shaping and legitimating society. In its origin as a body that the king would appear to, the public is by nature a responsive, reflexive, and thereby a responsible and empowered entity. Founded on the sovereign's need for approval during the contentious later years of the aristocracy (an approval that eventually was withdrawn), the public sphere served as a check on the State, a key force in civil society. In that respect, the public sphere served in the same capacity as media: at the same time that the newspaper, the gallery, the novel, the modern theater, music, and so on emerged, the public produced voices of criticism. And even if the equation of public space and public sphere was a tricky one, by understanding media as a space (or conversely space as a medium), it was nevertheless possible to draw a rough link between the two.

As many theorists observed, the twentieth century was witness to a long, sustained decline in the public sphere. In Habermas's analysis, this came about due to the contamination of the public sphere by private matters, most crucially its colonization by capital and the consequent transformation of the media from a space of discourse to a commodified realm. As media concentrated in huge conglomerates that were more interested in the marketing of consensus than in a theater of deliberation and had little use for genuinely divergent positions, mass media sought consensus in the middle ground, the political apparatus that Arthur Schlesinger called "The Vital Center."[26] The model of the public became one-way, the culture industry and the political machine expecting approval or, at most, dissent within a carefully circumscribed set of choices.[27]

Public space was not left unmolested. On the contrary, it was privatized, thoroughly colonized by capital, less a place of display for the citizen and more a theater of consumption under high security and total surveillance.[28] Under postmodernism the condition seemed virtually total, the public privatized, reduced to opinion surveys and demographics. If there was hope for the public sphere, it came in the form of identity politics, the increasing voices of counterpublics composed of subaltern peoples (in the developed world this would have been nonwhites, gays, feminists, youth, and so on), existing in tension with the dominant public. But if counterpublics could define and press their cases in their own spheres, for the broader public they were marginalized and marginalizing entities, defined by their position of exclusion.[29] Toward the end of postmodernism in the early 1990s, even identity politics became colonized, understood by marketers as another lifestyle choice among many.[30] But if this was the last capitulation of the old public as an uncommodified realm for discourse, it was also the birth of networked publics.

Today, we inhabit multiple overlapping networks, some composed of those very near and dear to us, others at varying degrees of physical remove. The former of these networks are private and personal, extensions of intimate space that are incapable of forming into networked publics. Instead, interest communities, forums, newsgroups, blogs, and so on are sites for individuals who are generally not on intimate terms to encounter others in public. As we have described throughout the book, these networked publics are not mere consumers. On the contrary, today political commentary and cultural criticism are as much generated from below as from above. From the deposal of Trent Lott to Rathergate, networked publics have drawn attention to issues that traditional media outlets missed or were reluctant to tackle.

The idealized model for networked publics is, as Yochai Benkler suggests, that of a "distributed architecture with multidirectional connections among all nodes in the networked information environment."[31] This vision of the network, commonly held as a political ideal for networked publics and sometimes misunderstood as the actual structure on which the Internet is based, is taken from RAND researcher Paul Baran's famous model of the distributed network. Where centralized networks are dominated by one node to which all others are connected, and decentralized networks are dominated by a few key nodes in a hub and spoke network, under the distributed model, each node is equal to all others.[32] Baran's diagram has been taken up as a foundation myth for the Internet, but not only was Baran's network never the basis for the Internet's topology, it bears little resemblance to the way networked publics are organized. Benkler points out that the distributed model is merely ideal, and if we seek a networked public sphere with everyone a pamphleteer, we will be disappointed. Networked publics are by no means purely democratic spaces in which every voice can be heard. That would be cacophony. But, Benkler continues, if we compare our current condition to the mass media of the 1990s and earlier as a baseline instead, we can observe real changes. Barriers for entry into the public sphere *have* been greatly reduced. It *is* possible for an individual or group of individuals to put out a message that could be heard globally with relatively little expense.[33]

Still there are very real threats to the networked public sphere, and Benkler, like many other theorists, warns of them.[34] In terms of infrastructure, the decentralized, not distributed, structure of the Internet allows governments, like China, to censor information they deem inappropriate for public consumption and the United States's National Security Agency (NSA) to monitor private Internet traffic. So far, networked publics have found ways of routing around such damage, like providing ways of getting around China's censorship and exposing the NSA's infamous room at the AT&T switching station in San Francisco.[35]

But centralization that would emerge from within networked publics is also a danger. Manuel de Landa observes that networks do not remain stable but, rather, go through different states as they evolve.[36] Decentralized and distributed models give rise to centralized models, and vice versa, as they grow. The emergence of networked publics just as mass media seemed dominant is a case in point. In his work on blog readership, Clay Shirky notes that diversity plus freedom of choice results in a power-law distribution. Thus, a small number of A-list bloggers attracts the majority of the readers. If tag-oriented search engines like Technorati or del.icio.us attempt to steer readers into the long tail of readership, they also reinforce the A-list by making evident the number of inbound links to any particular site.[37] Moreover, even if such sites, together with Google, YouTube, Netflix, iTunes, and other search engines, successfully redirect us to the long tail, together they form an A-list of the big aggregators. For now most of these are catholic in what content they include, but it is entirely possible this may change.

The long tail may prove to be a problem for another reason, what Robert Putnam calls "cyberbalkanization."[38] Given the vast number of possible clusters one can associate with, it becomes easy to find a comfortable niche with people just like oneself, among other individuals whose views merely reinforce one's own. If the Internet is hardly responsible for this condition, it still can exacerbate it, giving us the illusion that we are connecting with others. Through portals like Google News or My Yahoo and, even more so, through RSS (Really Simple Syndication) readers, Nicholas Negroponte's vision of the "Daily Me," a personalized newspaper freshly constructed for us every morning and tailored to our interests, is a reality. Even big media, under pressures of postfordist flexible consumption, has fragmented into a myriad of channels. But this desire for relevance is dangerous. It is entirely possible to essentially fabricate the outside world, reducing it to a projection of oneself. Rather than fostering deliberation, blogs can simply reinforce opinions between like-minded individuals. Conservatives talk to conservatives while liberals talk to liberals. Lacking a common platform for deliberation, they reinforce existing differences. Moreover, new divisions occur. Humans are able to maintain only a finite number of connections, and as we connect with others at a distance who are more like us, we are likely to disconnect with others in our community who are less like us. Filters too can lead to grotesque misrepresentations of the world, as in the case of happynews.com ("Real News. Compelling Stories. Always Positive.").

Another salient aspect of network culture is the massive growth of nonmarket production. Led by free, open-source software such as the Linux operating system (run by 25 percent of servers) and the Apache Web server (run by 68 percent of all Web sites), nonmarket production increasingly challenges the idea that production must inevitably be based on capital. Produced by thousands of programmers who band together to create software that is freely distributed and easily modifiable, nonmarket products are increasingly viable as competitors to highly capitalized products by large corporations.[39] Similarly, as our chapter on the topic points out, cultural products are increasingly being made by amateurs pursuing such production for networked audiences. Sometimes producers intend such works to short-circuit traditional culture markets, speeding their entry into the marketplace or getting past barriers of entry. At other times, such as in the vast Wikipedia project, however, producers take on projects to attain social status or simply for the love of it. Often these producers believe in the importance of the free circulation of knowledge outside of the market, giving away the rights to free reproduction through licensing such as Creative Commons and making their work freely accessible on the Internet. But P2P production also faces challenges. Chief among these is new legislation by existing media conglomerates aiming to extend the scope of their copyright and prevent the creation of derivative work. Even if advocates of the free circulation of cultural goods are successful in challenging big media, it is still unclear if the burgeoning fan culture is critical or if it only reinscribes, to a degree that Guy Debord could not have envisioned, the colonization of everyday life by capital, with debates about resistance replaced by debates about how to remix objects of consumption. Furthermore, the possibility of consumers not only consuming media but producing it for the (new) media outlets suggests the possibility of new, hitherto unanticipated, forms of exploitation.

By no means are network culture and the network economy limited to the developed world. If in this book we have largely looked at the most developed parts of the world, it is the consequence of our own individual biases, upbringings, and fields of study. Network culture envelops the entire world. If imperialist capitalism used the developing world for its resources and hand labor, and late capitalism exported manufacturing, networked capital exports intellectual labor and services.

But outsourcing is only a start. The mobile phone has revolutionized communication in the developing world, often leapfrogging existing structures. Due to the absence of any state apparatus that might regulate its phone system, Somalia, for example, has the most competitive communication market in Africa.[40] Nor is innovation in the developing world likely to cease. The developed world has only lukewarmly adopted mobile phones as platforms for connecting to the Internet, but for the majority of the world's inhabitants living in the developing world, such devices are likely to be the first means by which they will encounter the Internet.[41] History suggests that as different societies pass through similar levels of economic development at different times, unique cultural conditions emerge (e.g., Britain, the first country to industrialize, developed the Arts and Crafts movement, and some fifty years later

Germans responded to industrialization with the Deutscher Werkbund). The non-English-speaking developing world's reshaping of the Internet through the mobile phone will almost certainly be utterly unlike what we have experienced here.

All too often, discussions of contemporary society are depicted in the rosiest of terms. Sometimes this relentless optimism is a product of fatigue with outmoded models of criticism; sometimes this is just industry propaganda. But to be sure, network culture is not without its flaws. Many of these are nothing new, mere extrapolations of earlier conditions. As with modernism, and postmodernism before it, network culture is the superstructural effect of a new wave of capital expansion around the globe, and with it comes the usual rise in military conflict. Today's new wars are network wars, with networked soldiers and unmanned search-and-destroy flying drones fighting networked guerillas in what Castells once dubbed the "black holes of marginality," spaces left outside the dominant network but increasingly organized by networks of their own.[42] Closer to home, as Deleuze points out, the subtler, modulated forms of control in network culture mask themselves, above all in the idea that resistance is outmoded. This position, which Richard Barbrook and Andy Cameron have dubbed the "Californian Ideology," suggests that technology is inherently liberatory and that the network is both a space of self-realization and a natural road to a greater democratic government.[43] Under network culture, the idea that the corporation has a soul (which Deleuze declared "the most terrifying news in the world") and that the primary route by which individuals can achieve self-realization is through work, is commonplace, if perhaps treated with a little more skepticism since the collapse of the dot-com boom.[44] Moreover, as we explore the long tail, we are tracked and traced relentlessly, and as we are monitored, Deleuze concludes, we wind up internalizing that process—so as to better monitor ourselves.

If we have largely looked toward the utopian, positive moment in network culture, we note new threats emerging as well. Sensing that their day is done and that the means of production are in our hands, many large media outlets are fighting to extend their power through legislation, especially through radical modifications of the copyright law to prolong its length and expand its scope. As far as aggregators go, for now, Google's motto is "don't be evil." Given the corporation's recent compromise with China, allowing the government to censor its search engine results, precisely what is evil and what is not may be murkier than we might hope.[45] Another danger comes from telecoms, some of which dearly miss the monopoly status once enjoyed by the former AT&T. They hope to find salvation by controlling the means of distribution, profiting from giving privilege to certain network streams over others. Meanwhile

RFIDs and the ever-growing digital trail of information that we leave behind suggest that in the near future our every action could be tracked, not just by the government but by anyone able to pay for that information as well. All the while, whether network culture plants the seeds of greater democratic participation and deliberation, or whether it will only be used to mobilize already like-minded individuals, remains an open question. The question we face at the dawn of network culture is whether we, the inhabitants of our networked publics, can reach across our microclustered worlds to coalesce into a force capable of understanding the condition we are in and produce positive change, preserving what is good about network culture and changing what is bad—or whether we are doomed only to dissipate into the network.

Notes

1. Although there is much to recommend Carlota Perez, *Technological Revolutions and Financial Capital: The Dynamics of Bubbles and Golden Ages* (Northampton, MA: E. Elgar, 2002), she does not make a distinction between network society and the information age. Similarly, see Tiziana Terranova, *Network Culture: Politics for the Information Age* (London: Pluto Press, 2004).

2. Charlie Gere, *Digital Culture* (London: Reaktion Books, 2002), 11.

3. Hiroko Tabuchi, "PCs Being Pushed Aside in Japan," *Yahoo! News,* November 4, 2007, http://www.newsvine.com/_news/2007/11/04/1072065-pcs-being-pushed-aside-in-japan.

4. Jim Fisher, "Poison Valley (Part 1): Is Worker's Health The Price We Pay for High-Tech Progress?" salon.com, July 30, 2001, http://archive.salon.com/tech/feature/2001/07/30/almaden1/; Fisher, "Poison Valley (Part 2): What New Cocktails of Toxic Chemicals are Brewing In The High Tech Industry's 'Clean Rooms'—And Will We Ever Know What Harm They Are Causing?" salon.com, July 31, 2001, http://archive.salon.com/tech/feature/2001/07/31/almaden2/index.html.

5. Sassen, *The Global City: New York, London, Tokyo,* second edition (Princeton: Princeton University Press, 2001); Castells, *The Rise of the Network Society,* second edition (New York: Blackwell Publishers, 2000); Michael Hardt and Antonio Negri, *Empire* (Durham: Duke University Press, 2000); Castells, *The Internet Galaxy: Reflections on the Internet, Business, and Society* (Oxford: Oxford University Press, 2001).

6. Castells, *The Rise of the Network Society,* 500–509.

7. Jameson, "Postmodernism, or the Cultural Logic of Late Capitalism," *New Left Review,* 146 (1984): 53–92, later republished in expanded form as Jameson, *Postmodernism, or, the Cultural Logic of Capitalism* (Durham: Duke University Press, 1991).

8. Mandel, *Late Capitalism* (Verso: London, 1978).

9. Michael Hardt and Antonio Negri, *Empire* (Cambridge, MA: Harvard University Press, 2000).

10. On the anticipation of postmodernism by modernism, see Jameson, "Postmodernism," 56; Hal Foster, *The Return of the Real* (Cambridge, MA: MIT Press, 1996).

11. See Rosalind Krauss, *The Originality of the Avant-Garde and Other Modernist Myths* (Cambridge, MA: MIT Press, 1985).

12. Bourriaud, *Postproduction* (New York: Lukas & Sternberg, 2002).

13. Bourriaud, "Public Relations," interview by Bennett Simpson, *ArtForum,* (April 2001), 47.

14. By this I mean they tend to be done recently but can be taken from as far back as the early 1960s, when it had become clear that modernization, in its first phase at least, was complete and the idea of "the contemporary" began to emerge. Among the first cultural institutions to recognize this, the Museum of Contemporary Art, was founded in Chicago in 1967. On "the contemporary," see, for a start, Arthur Danto, *After the End of Art: Contemporary Art and the Pale of History* (Washington DC: National Gallery of Art, 1997), 10–11.

15. On nostalgia in postmodernism, see Jameson, "Postmodernism," 67. On allegory see Craig Owens, "The Allegorical Impulse: Toward a Theory of Postmodernism," parts 1 and 2, *Beyond Recognition: Representation, Power, and Culture* (Berkeley: University of California Press, 1992), 52–87. On periodization and network culture see Kazys Varnelis, "Network Culture and Periodization," http://varnelis.net/blog/kazys/network_culture_and_periodization. T. J. Clark, *Farewell to an Idea: Episodes from a History of Modernism* (New Haven: Yale University Press, 1999), 3.

16. On the end of the distinctions between high and low, see John Seabrook, *Nobrow: The Culture of Marketing and the Marketing of Culture* (New York: Alfred A. Knopf, 2000).

17. Gibson, *Pattern Recognition* (New York: Putnam, 2003).

18. Ronald Grover and Cliff Edwards with Ian Rowley, "Game Wars," *Business Week,* Feb 28, 2005, 60. On games, see McKenzie Wark, *Gamer Theory* (Cambridge, MA: Harvard University Press, 2007); Alexander R. Galloway, *Gaming: Essays on Algorithmic Culture* (Minnesota: University of Minneapolis Press, 2006).

19. Castells, 22.

20. Gilles Deleuze, "Postscript on Control Societies," in *Negotiations: 1972–1990,* (New York: Columbia University Press, 1990), 177–182.

21. "Myspace Quiz: What Character from a Chick Flick are you?" http://quiz.myyearbook.com/myspace/TelevisionMovies/45951/What_character_from_a_chick_flick_are_you.html.

22. danah boyd, "Social Network Sites: Public, Private, or What?" The Knowledge Tree, http://kt.flexiblelearning.net.au/tkt2007/?page_id=28.

23. "Learning to Live with Big Brother," *The Economist,* September 27, 2007, http://www.economist.com/world/international/displaystory.cfm?story_id=9867324.

24. cf. Jacques Derrida, *Writing and Difference* (Chicago: University of Chicago Press, 1978).

25. Bruno Latour, *Reassembling the Social: An Introduction to Actor-Network-Theory* (Oxford: Oxford University Press, 2005); Bruno Latour and Peter Weibel, *Making Things Public: Atmospheres of Democracy* (Cambridge, MA: MIT Press, 2005); Patricia Pearson, "Are BlackBerry Users the New Smokers?" *USATODAY.com OPINION,* December 12, 2006, http://blogs.usatoday.com/oped/2006/12/post_27.html.

26. Schlesinger Jr., *The Vital Center: The Politics of Freedom* (Boston: Houghton Mifflin Company for The Riverside Press, Cambridge, 1949); Max Horkheimer and Theodor W. Adorno, *Dialectic of Enlightenment,* trans. John Cummings (New York: Continuum, 1991; first published in English translation, Herder and Herder, 1972; originally published in German as *Dialektik der Aufklärung,* Amsterdam: Querido, 1944).

27. On marketing during the 1960s, see for example, Thomas Frank, *The Conquest of Cool: Business Culture, Counterculture, and the Rise of Hip Consumerism,* (Chicago: University of Chicago Press, 1997).

28. The classic work here is Richard Sennett, *The Fall of Public Man* (New York: Alfred A. Knopf, 1976).

29. On counterpublics see Oskar Negt and Alexander Kluge, *Public Sphere and Experience: Toward an Analysis of the Bourgeois and Proletarian Public Sphere,* trans. Jamie Owen Daniel, Peter Labanyi, and Assenka Oksiloff (Minneapolis: University of Minnesota Press, 1993).

30. See for example, Steven Kates, *Twenty Million New Customers: Understanding Gay Men's Consumer Behavior* (Binghamton, NY: Haworth Press, 1998).

31. Benkler, *The Wealth of Networks: How Social Production Transforms Markets and Freedom* (New Haven: Yale University Press, 2006), 212.

32. Baran, On Distributed Communications, (technical report, RAND Corporation, 1964), Vol. 1, http://www.rand.org/pubs/research_memoranda/RM3420/index.html . For a discussion of Baran's model and the Internet see Varnelis, "The Centripetal City: Telecommunications, the Internet, and the Shaping of the Modern Urban Environment," *Cabinet Magazine 17,* Spring 2004/2005; Janet Abbate, *Inventing the Internet,* (Cambridge, MA: MIT Press, 1999).

33. Benkler, *The Wealth of Networks,* 215.

34. For example, Richard Rogers, *Information Politics on the Web* (Cambridge, MA: MIT Press, 2004).

35. See "Boing Boing's Guide to Defeating Censorship," http://www.boingboing.net/censorroute.html; Ryan Singel, "Whistle-Blower Outs NSA Spy Room," *Wired.com,* http://www.wired.com/science/discoveries/news/2006/04/70619.

36. De Landa, *A Thousand Years of Nonlinear History,* (New York: Zone Books, 1997).

37. Clay Shirky, "Power Laws, Weblogs, and Inequality," e-mail to Networks, Economics, and Culture mailing list, February 8, 2003, http://www.shirky.com/writings/powerlaw_weblog.html.

38. Robert Putnam, "The Other Pin Drops," *Inc.,* May 16, 2000, 79, http://www.inc.com/magazine/20000515/18987.html; Carl R. Sunstein, "Democracy and Filtering," *Communications of the ACM,* 47, no 12 (December 2004): 57–59.

39. Michel Bauwens, "The Political Economy of Peer Production," *CTHEORY,* http://www.ctheory.net/articles.aspx?id=499.

40. "Somalia Calling," *The Economist,* December 2, 2005, 95.

41. Michael Minges, "Mobile Internet for Developing Countries," (proceedings, Internet Society Conference, Stockholm, Sweden, 3–5 June 2001), http://www.isoc.org/inet2001/CD_proceedings/G53/mobilepaper2.htm.

42. Castells, *The Rise of the Network Society,* 410.

43. Richard Barbrook and Andy Cameron, "The Californian Ideology," http://www.hrc.wmin.ac.uk/theory-californianideology-main.html.

44. Deleuze, "Postscript on Control Societies," 181.

45. Josh McHugh, "Google vs. Evil," *Wired* 11.01, January 2003, http://www.wired.com/wired/archive/11.01/google.html.

Notes on Contributors

Walter Baer Visiting Fellow, Annenberg Research Network on International Communication, University of Southern California

François Bar Associate Professor, Annenberg School for Communication, University of Southern California

Anne Friedberg Professor and Chair, Division of Critical Studies, School of Cinematic Arts, University of Southern California

Shahram Ghandeharizadeh Associate Professor, Department of Computer Science, University of Southern California

Mizuko Ito Research Scientist, School of Cinematic Arts, University of Southern California, and a Visiting Associate Professor at Keio University in Japan

Mark E. Kann Professor of Political Science and History, holds the USC Associates Endowed Chair in Social Science, University of Southern California.

Merlyna Lim Assistant Professor, School of Justice and Social Inquiry & Consortium for Science, Policy and Outcomes, Arizona State University

Fernando Ordonez Assistant Professor, Department of Industrial and Systems Engineering, University of Southern California

Todd Richmond Senior Research Associate, Institute for Creative Technologies, University of Southern California

Adrienne Russell Assistant Professor, Digital Media Studies, School of Communication, University of Denver

Marc Tuters Artist and researcher in interactive media

Kazys Varnelis Director of the Network Architecture Lab, Graduate School of Architecture Preservation and Planning, Columbia University, and Founding member of the Faculty, School of Architecture, the University of Limerick, Ireland

Index